Waterfall

Brian J. Hudson

REAKTION BOOKS

Published by
Reaktion Books Ltd
33 Great Sutton Street
London EC1V 0DX, UK
www.reaktionbooks.co.uk

First published 2012

Printed and bound in China by Eurasia

British Library Cataloguing in Publication Data
    Hudson, Brian J. (Brian James)
    Waterfall : nature and culture. – (The Earth)
    1. Waterfalls. 2. Waterfalls in art. 3. Waterfalls – Mythology.
    I. Title II. Series
    551.4'84-dc23

ISBN 978 1 86189 918 7

WATERFALL

The Earth series traces the historical significance and cultural history of natural phenomena. Written by experts who are passionate about their subject, titles in the series bring together science, art, literature, mythology, religion and popular culture, exploring and explaining the planet we inhabit in new and exciting ways.

Series editor: Daniel Allen

In the same series
Volcano James Hamilton

CONTENTS

Preface

'All possible variety of cascades, water-falls, and cataracts ...'
Thomas West, 1784

Many people take pleasure in visiting waterfalls and much has been written on the subject. Numerous accounts of Niagara Falls were published after Hennepin's late seventeenth-century descriptions, particularly from the early nineteenth century, but it was only later that other waterfalls became the subject of books. George Holley's *Niagara and Other Famous Cataracts of the World*, published in 1883, and John Gibson's *Great Waterfalls, Cataracts, and Geysers*, published in 1887, are early examples of global accounts of major falls.

Most books about waterfalls are guides to the falls of a particular country, state or region. Apart from a few slim illustrated volumes, few books have been published on the world's waterfalls since Edward Rashleigh's *Among the Waterfalls of the World* (1935).[1] Most of these are slim pictorial volumes, some aimed at the children's market. Geologist Richard Maxwell Pearl published a series of waterfalls articles in his journal *Earth Science* between 1973 and 1975, apparently with the intention of turning them into a book, but this never materialized.

My book, the culmination of more than a decade of waterfalls research, is comprehensive in its approach, but is not intended to describe as many of the world's waterfalls as possible. This is far from my aim, and readers may be disappointed at my omission of falls they feel deserved mention. What I have attempted to do is celebrate the delights of these beautiful wonders of nature by considering them from many points of view, emphasizing the roles that they play in the human experience.

To be as representative as possible, I draw on examples of waterfalls from all over the world, some famous, many not. North and South America, Europe, Africa, Asia and Oceania and, with recent global warming, the Earth's polar regions, all feature in the discussion. Even though there are already enough books and articles about Niagara Falls to fill a large library, it has been impossible to avoid making frequent reference to this great cataract, which has been so important in the history of travel and tourism, power generation, urban development and art. Among the issues that I consider is the human impact on waterfalls, particularly the effects of hydropower schemes and tourism development. Also considered are artificial waterfalls, which have long been features of the designed landscape. Their contemporary role is poignantly exemplified in the design of the National September 11 Memorial, in which the footprints of the Twin Towers are traced by walls of waterfalls.

A geographer and urban and regional planner by training, I have ventured into many other fields of knowledge that are outside my areas of expertise. I apologize for any errors that I may have made in my book and invite correction.

1 Waterfall Lovers and Waterfalling

'Waterfalls have universal appeal . . .'
Griff Fellows, *The Waterfalls of England*[1]

'Good waterfalling!'
Mary Welsh, *More Walks to Yorkshire Waterfalls*[2]

Many people take delight in waterfalls. They are variously known as waterfall lovers, fans, buffs or even collectors, although all that these enthusiasts are likely to take and keep are photographs. Treasured memories may also be kept in travel notes, letters and diaries. Armed with cameras, or perhaps with sketchbooks and pencils, they may, like the poets Wordsworth and Coleridge, set out 'to hunt waterfalls', as Dorothy Wordsworth expressed it in her diary.[3] Today this activity is commonly referred to as 'waterfalling', and among enthusiasts the term 'waterfallology' has come into use to describe the study of or active interest in waterfalls.

Waterfall lovers come from all walks of life and can be found all over the world. Some rural communities value their local waterfalls as places of relaxation where they can bathe in the tumbling streams with their refreshing cascades and inviting rockpools, then rest on the banks. Today, villagers in the mountains behind Finschhafen in Papua New Guinea walk to a little waterfall to enjoy a natural jacuzzi massage reputed to have healing powers.

Nowadays, most waterfall lovers are people who just like to visit waterfalls while on holiday or on excursions into the countryside. Waterfalls are popular destinations for day trips by car, often involving a short walk along the way or, for more remote falls, arduous hikes. They are commonly among the beauty spots that are visited on commercially operated tours. Some waterfall lovers take their interest more seriously, however. Many spend much of their leisure time searching for waterfalls that they have not seen before and revisiting favourites with which they are

familiar, enjoying them under various conditions that change according to season and weather. Some of these people are keen amateur and professional photographers who have a particular liking for waterfalls as subjects for their pictures. Others are more systematic in their approach, exploring particular areas and recording the falls they discover there, keeping detailed records of their height, width, flow and other characteristics. There are several highly informative and authoritative waterfall websites among the hundreds on the Internet. Many more are collections of photographic images, often holiday snapshots and accounts of trips.

J.M.W. Turner, *Hardraw Force*, 1816–18, watercolour. This waterfall in the Yorkshire Dales has been popular with 'waterfall lovers' since the 18th century.

Waterfalls, cascades and cataracts

Our most common experience of falling water is rain, but although we may enjoy this natural phenomenon, it does not give us that special pleasure we experience when visiting a waterfall. A waterfall may be defined as a 'vertical or very steep descent in a watercourse', but how steep does it have to be to qualify? At what angle does it cease to be a waterslide or rapid and become a waterfall? Even on a vertical or overhanging rock face, water streaming or trickling down may appear to the observer to be insufficient in quantity to make it a waterfall. Even well-known named falls may not always convince the observer that the descending water constitutes what is properly called a waterfall. When nineteenth-century travellers Samuel Mossman and Thomas Banister visited Wentworth Falls in New South Wales, Australia, 'a famous waterfall on the Blue Mountains', they were 'so disappointed at the small volume of water it displayed' that they felt 'it was scarcely sufficient . . . to be considered

Thomas Heawood, *Fall of Weatherboard*, engraving; original drawing by John Skinner Prout (1805–1875).

a waterfall'.[4] Like many others, this much painted and photo-graphed cascade is often reduced to a trickle in dry weather and may even dry up entirely.

While some 'waterfallologists' debate the criteria for defining a waterfall, more interesting to me is the use of the words 'cas-cade' and 'cataract', which are often applied to falls of various kinds. Dictionary definitions help, but uncertainties remain. A cascade may mean 'a small waterfall' or 'a series of waterfalls'. It is not clear whether the falls in a cascade that is a series of water-falls have to be small. A cataract may be 'a precipitous waterfall' or 'a large waterfall'. One geographer suggests a more precise definition of a cataract, applying the term to 'a steep fall of less than 75° in a watercourse which may also include a number of small waterfalls. The total height of these waterfalls will be less than the total cataract height.'[5] This definition implies that a cataract is less precipitous than a waterfall, yet Niagara Falls and Victoria Falls, both extremely precipitous, are commonly des-cribed as cataracts. They are certainly large waterfalls in terms of width and volume, though not particularly high in world terms. The famous Cataracts of the Nile, on one of the world's greatest rivers, cannot be considered as waterfalls and are best described as rapids, stretches where the flow is especially fast and often broken by rocks. On another of Africa's great rivers there are what Edward Rashleigh dismissively described as 'the so-called falls of the Congo'.[6] Most notable among these are the Boyoma (previously the Stanley) Falls which, though more impressive than the Nile Cataracts, like the latter are not true waterfalls but rapids. In seven main descents they fall 60 metres over a 100 kilo-metre stretch of river. These, too, are often referred to as cataracts.

Very tall waterfalls tend to be relatively narrow and small in volume. They often descend in several leaps, thus making them cascades in one sense – 'a series of waterfalls' – even though falls of this kind can hardly be described as 'small', the other defini-tion. Yosemite Falls, for example, cascades 739 metres down its cliff face in three great leaps, and are generally regarded as one of the giants in the world of waterfalls. In my book, any vertical or very steep drop in the course of a river or stream can be

Multnomah Falls,
Oregon, descends 189
metres in two major
steps. Built in 1914, the
Benson Bridge affords
pedestrians spectacular
views from midway up
the waterfall.

Huka Falls on North Island, New Zealand, is noted for its powerful flow and brilliant blue colour.

termed a waterfall. Some falls may be interrupted by ledges or may descend in a series of steps, thus becoming a cascade, but I also use the latter term to apply to small waterfalls, particularly those where the volume is slight. To me, the word 'cascade' suggests a delicacy that may be displayed in falls which occur as a series of shallow steps or a waterfall that descends as a fine, transparent curtain. In contrast, I apply the word 'cataract' to falls that are huge in volume, tumbling tumultuously over a brink that may be wide though not necessarily very high. I use the word 'falls' both in a plural sense, referring to more than one waterfall, and as a singular term where one particular waterfall is being mentioned. This is common in the literature, as in the example, 'The Huka Falls is New Zealand's most visited natural attraction.' What it is that attracts us to waterfalls is the subject of a later chapter, but before considering this, it is appropriate to examine the origins of waterfalls, the way they evolve and the life in and around the tumbling streams.

opposite:
Huka Falls.

2 Waterfalls: Birth, Life and Death

'Waterfalls are significant items for geomorphic interpretation.'
O. D. von Engeln, *Geomorphology: Systematic and Regional*[1]

'Waterfalls are common, with white water spraying over the rounded black boulders . . . as it travels at break-neck speed it seems impossible that anything could live in it.'
David Boag, *The Living River*[2]

The global distribution of waterfalls

Waterfalls are unevenly distributed, rare or entirely absent over most of the earth's surface but common in some regions. This is because the conditions required for the formation of waterfalls are highly localized. While over 70 per cent of the earth's surface is covered with water, most of it is in the oceans and seas. Of the 2.8 per cent of the world's water found on land, most is in the form of ice at the poles and in glaciers. Less than one per cent of the world's water is fresh and of this, running water makes up only a tiny part. At any given time, as little as 0.0001 per cent of the total quantity of water on earth is in river channels, and the area that they occupy is only about a thousandth of the entire land surface.[3]

Waterfalls can occur only where there are marked variations in the altitude of land, so they are absent from flat, low-lying areas such as coastal and alluvial plains. They can occur in terrain that is generally level or gently undulating as long as it is sufficiently elevated and traversed by rivers that have eroded pronounced valleys or gorges in their downstream reaches, as at Niagara and Victoria Falls. Hilly or mountainous areas, plateau edges and some coasts that have suitable geological conditions or with particular geomorphological histories are most likely to have waterfalls, as long as there are rivers or streams. Arid regions normally do not have permanent waterfalls unless they are traversed by rivers from wetter regions, as in the case of Augrabies Falls on the Orange River, which sustains its flow when it enters the Kalahari Desert. Even in deserts, however, the occasional

Purlingbrook Falls, Queensland, one of the many waterfalls to be seen in Australia's Gold Coast hinterland.

rainstorm can generate spectacular ephemeral falls, such as those seen occasionally at Australia's Uluru, also known as Ayers Rock.

Many of the world's best-known waterfalls are found in North America and Europe. Switzerland and Norway have many famous examples, which is not surprising in view of their rugged mountains and ample rainfall and snowfall. Great Britain, too, despite its relatively modest hills and mountains, boasts many waterfalls whose fame is disproportionate to their physical scale. Surprisingly, for so dry a continent with such generally low relief, Australia has many waterfalls that are often included in lists of the world's highest. More surprising, perhaps, is the paucity of well-known examples from the Himalayas and Andes, the world's highest and most extensive mountain ranges. With the possible exception of Tequendama Falls in the Colombian Andes and the recently 'discovered' Gocta Falls in Peru, neither of the world's greatest mountain systems has a waterfall that is internationally very famous. Tequendama Falls are highly accessible, a mere 30 kilometres from Colombia's capital city, Bogotá, a fact that no doubt helps to explain the fame of this waterfall. The best known Himalayan waterfalls are those within easy reach of urban centres, particularly those which were developed for colonial administration purposes. At Shillong in India, for example, Crinoline Falls lie within the town, while several other named falls are nearby. None of these is particularly notable in world terms, but they would have appealed as picnic spots or as places suitable for excursions among the colonial elite. Today they are again being promoted as tourist attractions.

In contrast to the still largely unknown Himalayas and Andes, the comparatively small area of western New York State contains over a thousand recorded waterfalls. This number is due largely to the proximity of one of the world's most highly populated areas where many people have the leisure and wealth for recreational travel, and to the diligent work of Scott Ensminger, who is engaged in a detailed survey of the waterfalls of that region. Another important factor to be noted is that Ensminger defines a waterfall as an abrupt descent in a stream that is at least 152 cm in height, thus including falls that in many parts of the world would scarcely be noticed. All this clearly shows that existing lists and accounts of the world's waterfalls reflect not only where waterfalls occur but also where people have observed and recorded them. Waterfalls are more likely to be known in well-populated areas or ones that have been 'discovered' for tourism. As more remote parts of the world become easily accessible to tourists and are being developed for tourism, more waterfalls are becoming identified and increasingly promoted as scenic attractions. This is true of remote and formerly inaccessible mountain areas such as the Andes and Himalayas. Continuing exploration of these wild mountains has led to the discovery of waterfalls that were previously unknown to the wider world. In late 1998 a National Geographic Society exploration team eventually reached what they named Hidden Falls on Tibet's Tsangpo River. This 35-metre waterfall was previously known only to local hunters and monks, their reports treated with suspicion by Westerners. The world's highest mountains, long famed for their lofty snow-capped peaks and huge glaciers, are now becoming known in greater detail by trekkers and other tourists, the regions' numerous, often spectacular waterfalls becoming increasingly familiar to visitors, and publicized by tourism promoters.

As recently as 2006, the 'discovery' in Peru of the 'World's Third Highest Waterfall' was announced in the world's press. The 771-metre Gocta Falls in the Andes of Amazonas, 700 kilometres north-east of Lima, was known only to the local people until a group of explorers arrived there in 2002 and reported their discovery to the Peruvian government. This stimulated plans for tourism development in the area.

Another sign of the times, a consequence of global warming, is that waterfalls are on the increase in polar regions, where melting ice now feeds falls in places where formerly water remained in frozen stillness. Plunging into gaping crevasses, over cliffs of shelf ice or down rock faces recently exposed by the retreat of glaciers, streams of meltwater now form waterfalls in Antarctica and Greenland, which were until recently almost devoid of these landforms. Climate change can be expected to affect the distribution and regimes (variability of discharge) of waterfalls elsewhere in the world, with desiccation in some regions and increasing storm frequency in others being among the important influences.

How waterfalls are formed

Most waterfalls are formed by the erosive action of moving water, in its liquid form as rivers and streams or as breaking sea waves, or in solid form as glacial ice. Irregularities in the longitudinal profile of the stream tend to develop in the form of rapids and falls as the flowing water erodes the different beds of rock over which it passes, cutting down into the less resistant softer or more fractured rock more quickly than into the tougher beds. Vertical movements of the earth's crust and faulting may also be important factors in the development of waterfalls, as can a drop in sea level. On some coasts it is the erosive action of the sea that creates waterfalls where streams tumble over cliffs formed by wave action, while in areas that have experienced glaciation many falls are the result of gouging by moving ice. Most waterfalls are the result of erosion of some kind, but there are others that are the result of constructional processes. Examples include those created when a landslide, lava flow or glacial moraine obstructs a stream, the water level rising behind the natural dam then spilling over as a waterfall. Often, the obstruction causes a diversion, sometimes sending the displaced stream over a cliff elsewhere. In some limestone regions, waterfalls may form in consequence of the gradual deposition of calcareous material precipitated from flowing water, a chemical process that can create natural barriers of travertine or tufa across the stream bed. Many artificial waterfalls are formed by the construction of barriers across rivers and streams. These include those that overspill dams built across watercourses for purposes such as water supply, irrigation and power generation.

Once formed, natural waterfalls continue to be subject to the forces and processes that created them. In this way their form and position change over time. In rare cases, this occurs so rapidly as to be noticeable in a single lifetime. More often the changes are imperceptibly slow in human terms, though rapid on the geological time scale.

Thomas Moran, *Great Falls of Yellowstone*, 1891, oil on canvas. This view captures the rugged grandeur of a gorge still undergoing erosion by a powerful river.

Waterfalls are normally formed by the erosive action of flowing water in rivers and streams. In this fluvial process, the hydraulic force of water may be augmented by the battering and abrasive action of the rocks and stones that are moved along the bed of the stream by powerful currents, especially in flood conditions. This ability to erode the stream bed, cutting down into the rocks and deepening the valley, can be greatly increased by vertical earth movements that raise the land surface, or by a drop in sea level. To understand the effects of this on fluvial processes, it is necessary to recognize that, as a river erodes its bed, it develops a longitudinal profile which tends towards a smooth curve; this is relatively steep near the source, becoming progressively less so towards the mouth. Until this graded stream profile is achieved, irregularities in the river bed may occur, reflecting the nature of the bedrock, the history of earth movements in the area, or both. Waterfalls formed by fluvial erosion generally indicate geologically recent rapid uplift of the land surface or a fall in sea level. Either of these events has the effect of bringing the adjacent offshore area up above sea level. Normally, the offshore slope is steeper than the gradient of the lower courses of the streams that flow into sea. Consequently, near the new mouth of the river the flow of water is accelerated, increasing its erosive power. This initiates change in the longitudinal profile of the stream bed, and the place where an abrupt steepening of the longitudinal profile occurs as a result of this process is called a knickpoint. This commonly manifests itself as rapids or falls, and a series of these in the course of a river may reflect a sequence of vertical earth movements or falls in sea level.

While knickpoint development may cause rapids or waterfalls in areas of homogeneous rock, falls are most likely to develop where a stream flows over rocks of varying resistance to erosion. For example, a bed of hard solid rock is likely to be eroded more slowly than one that is soft or has been shattered by geological processes. Where beds of rocks of varying resistance to erosion are juxtaposed along the course of a river, differential erosion is likely to create irregularities in the stream bed, and this often leads to the development of rapids and waterfalls.

The arrangement of the beds of rock, as well as their physical and chemical qualities, is particularly influential in the way that falls develop. Where the beds are more or less horizontal, conditions may be suitable for the development of cap-rock falls. As the stream erodes its bed vertically, it cuts into the superimposed strata below, wearing away the less resistant rocks more rapidly than the more resistant ones. This process leads to the formation of waterfalls where a bed of hard rock overlies less resistant rocks. As the rocks at the foot of the waterfall are quickly eroded, the cliff face recedes. The more resistant beds of rock at the top of the fall are undercut, creating an overhang which, in time, collapses under the force of gravity. The continued erosion and collapse of the rock face causes the site of the fall to retreat, its migration upstream creating a gorge below. Chunks of the collapsed cap rock may accumulate at the foot of the fall, being removed by the stream flow only after a long period of weathering and erosion has broken it into smaller fragments. Where the volume and height of the fall are great enough, however, the power of the falling water and the swirling currents below may be sufficient not only to keep the base of the cataract clear of accumulated rock debris, but to erode a deep hollow in the river bed. This is known as a plunge pool.

Many accounts of waterfalls describe this process of fluvial erosion but ignore the role of weathering, which can be important in the shaping and upstream migration of falls. Weathering refers to the disintegration of rock by physical processes such as frost shattering or chemical processes such as solution, as distinct

This winter view of Niagara Falls shows the resistant cap rock in the top left of the photograph. Faintly visible through the spray are rocks that have fallen from the undercut cliff.

from erosion by moving water (rivers, sea waves), ice (glaciers) or air (wind).

While many waterfalls develop on horizontal or gently inclined strata, others occur on beds of rock that have been tilted or contorted by earth movements that commonly force rocks which were originally deposited horizontally into the vertical position. Additionally, in periods of volcanic activity, molten material from deep within the earth is intruded into the rocks above. These igneous intrusions often solidify to form 'dykes', long vertical sheets of rock that become exposed on the earth's surface as a result of erosion. Where a stream flows across vertical beds of rocks, a particularly resistant bed may act as a barrier to erosion while softer rock downstream is more readily worn away. This can form what is known as a 'vertical barrier fall'.

Juxtaposition of rocks that are very different from each other is often brought about by earth movements along geological faults. When these rocks are markedly different in terms of resistance to weathering and erosion, waterfalls may develop where streams cross the fault line. It is important to recognize that falls such as these are usually caused by the differences in the rocks brought into contiguity by faulting, not by the deformation of the land surface due to earth movements. Falls are sometimes created in this way but, depending on the local geology, rapid erosion may cause them to disappear quickly.

While most textbooks emphasize the role of undercutting of cap rock in the formation and retreat of waterfalls, many actively retreating waterfalls are not undercut, but are buttressed outwards at their base. The explanation relates to the question of stress in a vertical rock face, the internal pressures and tensions that can cause collapse. Vertical and overhanging cliffs are inherently less stable than those which are buttressed at the base. Also important are the physical and chemical properties of the rocks and the geological structure, including the angle of inclination of the strata and the degree of jointing. Then there are the various ways in which water acts upon the rock, not only the erosive forces of the flowing river, but also the effects of groundwater within the strata.

Here seen from the Zambian side, Victoria Falls tumbles into a gorge created by fluvial erosion along fissures in the basalt Batoka Plateau.

So far, the discussion has focused on waterfalls that have been formed by the action of the streams on which they occur. In many cases, however, falls are caused by other geomorphological occurrences that can impinge on a stream: marine erosion, glaciation and volcanic activity, for example. Commonly, it is the work of another, larger, river that produces the fall in a tributary stream.

Where the erosive power of the main stream enables it to cut down rapidly into the bed rock, a small tributary may not be able to match this pace of erosion. This can lead to the development of a discordant junction of tributary and trunk stream, the latter flowing at a much lower level than the mouth of the watercourse entering it. In consequence, the tributary joins the main stream by a steep descent, and many waterfalls have been formed in this way. The explanation generally lies in the much smaller volume of water and the relatively little abrasive sediment that the tributary stream contains.

The development of discordant tributary valleys is somewhat similar to the formation of hanging valleys by the erosive action of glaciers in mountain areas. Active ice sheets and glaciers, particularly when armed with embedded rock fragments, have the power to gouge, pluck and scour the toughest rocks, but the effectiveness of glacial erosion is still influenced by variations in the resistance of different kinds of rock over which the ice moves. Consequently, when the ice disappears with climatic change, the newly exposed glaciated terrain is likely to exhibit an irregular surface with a disrupted drainage pattern, characterized by numerous lakes connected by rivers which tumble over rapids and falls. This kind of landscape is typical of much of Canada, Sweden and Finland, where the last Ice Age ended less than 10,000 years ago. In mountain regions where glaciers occupy valleys originally formed by fluvial erosion, the action of the moving ice modifies the valley floor, differential erosion often giving it an irregular, sometimes stepped longitudinal profile. This is the origin of many waterfalls in the glaciated valleys. The deepening and widening of a valley by glacial erosion also modifies the cross profile. The original valley, typically V-shaped, is thus converted into the U-shaped trough commonly associated with glaciated highlands. Smaller glaciers joining the main one from tributary valleys have less power to erode, and discordant junctions are normally created in this way. The consequences resemble those of fluvial downcutting by rivers, likewise forming hanging valleys from which tumble waterfalls.

Stoney Creek Falls, on a tributary of Queensland's Barron River, cascades into the gorge formed by the more powerful main stream. This is a regular stop on the Kuranda Scenic Railway.

Thomas Hill, *Bridal Veil Falls, Yosemite Valley*, *c.* 1880s, oil on canvas. This waterfall drops from a hanging valley into a glacially gouged trough.

26

Milford Sound, New Zealand, is a fjord, a glacially eroded coastal valley that has been inundated by the sea. Spectacular waterfalls cascade down its sides.

On some glaciated rocky coasts, marine falls or rapids occur where sea inlets have rock barriers that interfere with the ebb and flow of the tides. This phenomenon is typically associated with a barely submerged ridge of rock extending across the inlet. At low tide, water falls over the rock barrier as it flows towards the sea; on the flowing tide, the rising seawater passes over the same rock formation, but in the opposite direction. Scotland's Falls of Laura is a notable example. Canada's more famous Reversing Falls are mainly the consequence of exceptionally high tides and a narrow stretch in the channel.

By no means all coastal waterfalls are of glacial origin. On reaching the coast, a stream always tends to erode its bed down to sea level, or to raise it by deposition when sea level is on the rise. On many rocky coasts, marine erosion is causing the retreat of cliffs at a rate that small streams cannot match as they erode their beds vertically. Unable to cut down fast enough to maintain a gentle gradient to the sea, at its mouth the stream dashes over a cliff formed by the erosive action of breaking waves. At

A coastal waterfall at Runswick Bay, North Yorkshire. At high tide the foreshore is covered by the sea, which continues to erode the cliff.

The stream that drains Loch Mealt on the Isle of Skye plunges vertically over a sea cliff formed by wave action.

low water, the stream may run its last few metres across the foreshore to reach the sea, but at high tide it will plunge straight into the waves that continue to cut back the cliffs.

So far in this chapter, all the waterfalls that have been discussed are products of erosion. There are some falls, however, that are caused by the obstruction of the stream channel. The most common of these are the artificial falls which cascade over dams when the reservoirs behind them are full, but streams are sometime dammed by natural processes. Landslides, lava flows and glacial moraines are among the natural dams that can cause waterfalls. Because of the unconsolidated nature of the material vthat forms the obstruction, waterfalls created in this way are usually short-lived, the stream rapidly eroding a channel through the barrier in its path. In many cases, falls that are caused by obstructions of this kind occur not at the site of the natural dam, but on a new course of the stream diverted by the blockage.

A relatively rare type of waterfall is the kind which forms where the stream itself constructs a barrier over which it cascades. This occurs in some limestone areas where cascading streams are often characterized by deposits of calcite known as travertine or tufa, depending on their precise composition. Impressive examples are found in the Plitvice Lakes National Park, Croatia. Here travertine barriers on the Korana River create a series of waterfalls and cascades, some up to 80 metres high. The Plitvice region has dramatic scenery, both above and below ground, and caves are common there.

The chemical and physical characteristics of some limestones often produce very distinctive landscape features known collectively as karst topography. Subterranean drainage systems are common in such areas, and waterfalls frequently occur on the underground streams. The formation of the interconnected underground passages and caves found in karst areas is due mainly to the solubility of calcium carbonate, the main component of limestone, in rainwater. The water, having dissolved carbon dioxide from the atmosphere, is weakly acidic. When it percolates down through the limestone, it reacts with the calcium carbonate and removes it in solution. This creates voids in the rock, underground spaces through which substantial streams may flow. Thus are added the effects of fluvial erosion to those of the chemical process of solution in the development of subterranean scenery. Cave systems develop best where the limestone is characterized by cracks in the rock, known as joints. Water moves more easily along the joints that are thus widened by solution, forming subterranean passages and caves. Vertical joints facilitate the downward movement of water, contributing to the development of shafts down which waterfalls may tumble. Streams sometimes enter limestone cave systems by falling into swallow holes, depressions on the surface leading to vertical shafts such as that at Gaping Ghyll in North Yorkshire, where England's (but not Britain's) highest waterfall drops more than 100 metres into a large cavern.

Gaping Ghyll in the Yorkshire Dales: a stream plunges underground, forming a subterranean waterfall.

Before leaving the subject of sub-surface waterfalls, mention should be made of ocean cataracts. These are enormous flows of colder, and therefore denser, water that descend to great depths on the margins of several oceanic basins. They are, in effect, waterfalls within the oceans themselves. That which occurs in the Denmark Strait, between Greenland and Iceland, is about 3.5 kilometres in height, more than three times higher than Angel Falls, which are generally regarded as the world's tallest. This submarine 'waterfall' has a flow rate of five million cubic metres of water per second and may be considered as the world's greatest cataract. Not all ocean cataracts are generated by temperature differences. That which flows out of the Strait of Gibraltar into the North Atlantic is driven by a difference in salinity.

Artificial waterfalls

One of Europe's grandest and most famous waterfalls, the Cascata delle Marmore near Terni, Italy, is the result of diversion, but in this case the geomorphological agent was human. These falls were created in 271 BC as a consequence of a Roman civil engineering project. This was undertaken as a flood mitigation and drainage measure involving the diversion of water from the sluggish Velino River into the Nera, the new course passing over a cliff to form a spectacular waterfall. During the following two millennia several major modifications were made to this river diversion scheme. The 165-metre-high Cascata delle Marmore was long admired as one of Europe's most impressive waterfalls, but since being harnessed for hydroelectric power generation its flow has been greatly diminished. To cater for the demands of tourism, the falls are 'switched on' at advertised times, when the full flow of the river is allowed to pass over the brink.

Even 'natural' waterfalls are not always what they seem. Apart from the turning on and off of flows, there are other ways in which owners and managers of waterfalls interfere with the processes of nature and natural landscapes. Niagara Falls, for example, has been much modified by human action, and not only by engineering works associated with the hydroelectric power development

One of Europe's most impressive waterfalls, Cascata delle Marmore, near Terni, Italy, is a result of a Roman engineering scheme.

and the consequent reduction of flow over the brink. Among the most notable alterations was the destruction, in 1955, of the famous Cave of Winds where formerly visitors could walk behind the curtain of falling water. This was done as a safety measure after the deaths of several tourists at this spot.

Tourism is an important factor in the creation of artificial waterfalls, which are commonly found in resorts and theme parks. A theme park at Shenzhen, China, features scale models of some of the world's most famous tourist sights, like the Pyramids, the Taj Mahal, the Eiffel Tower and Niagara Falls. In Singapore, where waterfalls are not features that most visitors would expect to see, the artificial Jurong Falls appear remarkably natural in the rainforest setting of an open access aviary in Jurong Bird Park. Here recirculated water plunges over a 30-metre cliff amid lush tropical vegetation.

35

Artificial waterfall at Banishead Quarry, Cumbria.

For many years the role of human activity in shaping the earth's surface was largely neglected by geomorphologists, but since the 1960s considerable attention has been given to artificial or anthropogenic landforms. These include features that are produced by excavation, which may be considered a form of erosion, and deposition, the latter including the dumping of earth or rubbish and construction work of various kinds. This usually involves the use of tools and machines including earth-moving equipment. 'Bulldozogenesis', a word humorously coined in the 1960s, is an apt term for the process that shapes much of the contemporary landscape. An artificial waterfall can be formed by the excavation in the path of a watercourse. Abandoned Banishead Quarry, with its artificially created fall, is now an admired scenic attraction in the Lake District.

Found throughout much of the world are the artificial falls created where dams and weirs are built across rivers and streams to store and divert water for various purposes, including irrigation, potable water supply and power generation. When the amount of water behind the dam is more than is needed, it is usually allowed to spill over, creating a waterfall that can be an impressive sight.

The death of waterfalls

All landforms are destined to disappear eventually, and in terms of geological time waterfalls are amongst the most ephemeral. Waterfalls, however formed, are inherently self-destructive landforms. Whether by cutting back or cutting down or both, fluvial erosion, assisted by weathering, will gradually remove these irregularities in the longitudinal profile of the stream, tending to produce one that displays the typical smooth curve from source to mouth. Renewed uplift of land areas or falls in sea level may interrupt this process, however. Repeated changes of this kind can renew the energy and erosive power of a stream which, thus continually rejuvenated, creates new waterfalls as it also destroys the older ones retreating upstream.

Many waterfalls, now vanished, have existed in the past, long before there were human eyes to see them. Many falls, which our human ancestors may have seen, have also disappeared, some leaving little or no trace, others looking almost as they did when still active, lacking only the descending water. Probably the most dramatic of these is Dry Falls, on the Columbia Plateau of the state of Washington. Today Dry Falls, near Coulee City, is a 5.5-kilometre horseshoe of cliffs 120 metres high. Several thousand years ago, towards the end of the last Ice Age, this was the site of an enormous cataract with an estimated discharge some forty times that of Niagara. The cause of this phenomenon was the break-up of a natural dam of ice that held back a huge glacial lake in what is now western Montana, some 10,000 to 13,000 years ago. This occurred repeatedly, releasing vast quantities of water that eroded channels in the Columbia Plateau and created many waterfalls, including that now known as Dry Falls. Here the plunge pools survive to this day, thousands of years after the falls ceased to flow, testimony to the erosive power of the water that formerly tumbled over the cliffs.

Marine cataracts on a vastly greater scale than the waterfalls we can see today have probably existed in the distant past, falls different not only in size but in their catastrophic irreversibility. There is evidence that five and a half million years ago the rising

waters of the Atlantic Ocean broke through the ridge that formerly linked Africa with Europe where the Strait of Gibraltar is today. The basin now occupied by the Mediterranean Sea was then a desert, an earlier sea there having dried up, leaving only a few brackish lakes. With a rise in ocean levels fed by melting ice as an Ice Age ended, the North Atlantic breached the barrier separating it from the Mediterranean Basin. The unimaginably huge cataract may have taken a century to fill the Basin.[4]

Dry Falls, Washington state.

The historical significance of a similar occurrence is dramatically illustrated by recent research that shows that between 450,000 and 200,000 years ago the waters from a lake created by an ice sheet that occupied the present site of the North Sea spilled over the ridge that formerly connected what is now Britain with the rest of Europe. With a discharge 100,000 times greater that the Thames, the torrential floodwater carved the breach that we now know as the Straits of Dover. Surveys of the sea floor of the English Channel have revealed submerged features which are evidence of cataracts up to 30 metres high that have cut down and back into the rock, contributing to the process of erosion that eventually made Britain an island.[5]

During the millions of years of the restless earth's geological history, there is no doubt that many catastrophic events of this kind have occurred. It seems probable that the constant rise and fall of different parts of the earth's crust and the changing levels of seas and oceans will, from time to time, produce conditions similar to those that created Gibraltar Strait Falls and the English Channel. Meanwhile, the geomorphological processes that continually operate on and under the earth's surface create and destroy waterfalls over periods commonly measured in thousands or tens of thousands of years.

During the life of a waterfall, its appearance may change considerably. The visual form that a waterfall takes depends on several factors, including local geology, past and present processes that have shaped the fall, and the stage of development. Waterfalls have been classified according to the forms they take, and although writers are not in complete agreement on terms and definitions, the following will be helpful. A *block* is a relatively wide fall, broader than it is high. A *curtain* is also a relatively wide fall, but higher than it is broad. A *fan* remains in contact with the bedrock, with the falling water increasing in width as it descends. A *punchbowl* is a narrowly constricted fall that drops into a rock-pool. A *horsetail* maintains some contact with the bedrock, the water falling without fanning out much, thus resembling the tail of a horse. A *plunge* is a vertical fall that loses contact with the bedrock. A *segmented* fall splits into several channels in its descent. A *tiered* fall descends in several distinct steps. A *multi-step* is a series of falls, each with its own plunge pool.

Waterfalls and the natural environment

Although rivers and streams account for a very small fraction of the earth's surface and waterfalls are highly localized occurrences on them, the ecological importance of these freshwater bodies and landforms is enormous. Whether huge rivers or tiny rills, watercourses are essential for life far beyond their immediate

This Guyana Tourism Authority poster promotes the spectacular Kaieteur Falls, together with the rich flora and fauna found in the vicinity of the cataract.

environs, and are visited by creatures of all kinds that need water to drink in order to survive. Some travel great distances to slake their thirst. In and around freshwater bodies, distinctive ecosystems flourish, and where waterfalls occur, a variety of plant and animal species have adapted to the challenging conditions presented by falling water, rapids, powerful currents and turbulence. Waterfalls also have a great influence as barriers to movement of life forms up and down watercourses.

Large waterfalls act as barriers that may interrupt coloni-zation by some species. There is evidence that waterfalls can so isolate the headwaters of rivers that distinct varieties or even unique species of fish may evolve as a result.[6] Waterfalls commonly form obstacles to migration but, as the well-known example of salmon testifies, some creatures are able to surmount low falls and rapids, particularly where the descent is broken into separate cascades forming a natural staircase. While no fish can swim against falling water, many jump effectively. Atlantic salmon can clear over three metres. To enable it to jump higher, the salmon leaps from the crest of the standing wave that commonly forms near the foot of a fall. Here there is an upward component in the turbulence which assists the fish in its leap. To facilitate the movement of fish upstream, fish passes, artificial stepped channels bypassing falls, have been built on many rivers, while in some places the falls themselves have been deliberately modified or destroyed for this purpose.

When descending rivers, fish may have no difficulty negotiating falls and rapids, particularly where the drop is small enough for them or their parents to have ascended. Here the cushioning effect of the water is sufficient to prevent damage. Waterfalls can be destructive of aquatic life, however. Researchers have found that many species of plankton, minute water animals and plants usually living in colonies, are eliminated by falls and rapids. Others have shown that some species can survive the descent of steep rapids and even major cataracts.

Unfortunately for migrating salmon, there are often other hazards at waterfalls in the form of preying animals, including humans, that are attracted to these places by the prospect of a large and relatively easy catch when the fish are especially vulnerable. Among the animals that come to waterfalls to feast on salmon in the spawning season are the black and brown bears of North America. The fishing exploits of these large animals in the tumbling rivers have become a popular tourist attraction in recent years.

Swimming and jumping are not the only means by which fish can surmount the barriers imposed by rapids and waterfalls. Species with suckers or friction discs can often ascend wet vertical rock faces beside or beneath falling water in stream courses.

Some fishes belonging to the large family of gobies (*gobiidae*) have the ability to climb waterfalls in this way. Other fish that have the ability to overcome waterfall barriers include young eels, which make use of wet areas beside the stream to bypass falling water.

While falls often act as barriers to movement and form obstacles in the stream that separate reaches where life can be readily sustained, they provide a habitat to which some species are well adapted. There are plants and other organisms that survive and even flourish in the fast-flowing, turbulent water of falls and rapids. Its waters well-oxygenated by numerous cascades, the tumbling stream provides an ideal environment for primary production in the form of plants, algae and epiphytes, so long as there is adequate sunlight for photosynthesis. Particularly important are the small filamentous plants which grow closely attached to other plants, rocks, stones and dead branches in the stream. These form 'a dense, slimy felt of algae or mosses, which are the most important primary producers in running streams'.[7] The high level of productivity is due largely to the rapid current. This prevents the development of a film of nutrient- and carbon dioxide-depleted water around the plants that would reduce primary production. Within the mass of algae and epiphytes lives a very rich fauna of small animals, such as watermites, insect larvae, nematodes, water spiders and small crustaceans. Clinging to solid holds in the tumbling torrent and in the splash zones of waterfalls, these communities of plants and tiny animals flourish, providing food for the rest of the inhabitants of the stream. Many of the latter, however, are unable to live in the fast-moving, turbulent waters of rapids and falls, except for some species that pass through them for short periods on their way to more favourable environments. Indeed, the wildlife that is commonly found in and around torrential streams and waterfalls normally takes advantage of any available protection from the powerful currents, finding refuge behind or beneath boulders, or in gentle eddies and quiet pools outside the main flow.

In addition to behavioural responses such as these, some animals and plants display morphological and physiological adaptations to the environment. Mention has already been made of fish with anatomical characteristics that enable them to use friction and suction to ascend waterfalls, but there is a fascinating variety of ways in which plants and animals maintain holds and avoid being swept away in the rushing, falling turbulent waters. Some water plants, the Podostemons, maintain their hold by chemical adhesion to rock surfaces, the root system being effectively absent. They have also adapted to survive great fluctuations in water levels, including periods of low water when they dry out. By these means, they are successful in colonizing waterfalls and rapids. Found mainly in the tropics, they are the only flowering plants that normally grow in such places.

Creatures living in torrential habitats, such as riffle beetles (both adults and larvae), and the larvae of blackfly and some mayfly and caddisfly, commonly have a flattened body which appears to be a hydrodynamic adaptation to the environment. Some of the flattest of the animals found in streams are water pennies, which are the larvae of the riffle beetle family. These seem to adhere to the substrate through suction, and many different kinds of animals that live in and around waterfalls and rapids have various forms of suckers enabling them to maintain a hold in the rushing water. Others have claws or hooks to attach them to appropriate surfaces in the watercourse, while extruded silk is used for attachment and as lifelines by many insect larvae and pupae.

An Australian species associated with waterfalls is the waterfall frog (*Litoria nannotis*), a native of tropical north Queensland. This amphibian displays a range of adaptive characteristics that vary with the stage in its life cycle. Like some other animals that frequent waterfalls, the waterfall frog makes use of torrential streams as places of refuge, entering them when disturbed and hiding in crevices or under rocks beneath the water. It is also adapted morphologically and physiologically to this rigorous environment. The males have small spines on their thumbs and chest to help avoid being dislodged by the force of water during mating, while the tadpoles have sucker mouths so that they can attach themselves to slippery stones. The females lay eggs under

A waterfall-dwelling frog.

stones in a sticky mass that helps hold them in place. Another adaptation to the environment is the male's mating call, which is a growl instead of a more complicated call that would not be heard against the sound of falling water. A species of frog that lives behind a waterfall in South America has dispensed with vocal calls altogether, so noisy is its environment. Instead of making a futile attempt to compete with the loud sound of the falls, the male makes visual signals with its bright blue foot in order to communicate to females its desire to mate.

Among the many different creatures that can be found in waterfall habitats are birds of various kinds. These include the dippers, of which there are six species distributed widely across the globe, and a similar number of duck species. Dippers, especially, are strongly associated with waterfalls, often building their nests behind the curtain of falling water. Their morphological and physiological adaptations to their environment include dense, waterproof plumage and strong bills, legs, toes and claws. Their modified wings act like flippers, and they appear to fly under water and walk along the stream bed while foraging for aquatic insects, including the larvae of mayflies and caddisflies, together with some crustaceans, molluscs and fish. Unlike their cousins, the two Andean dipper species do not dive or swim: instead they catch prey from rocks in cascades and waterfalls or by wading into shallows and dipping their heads beneath the water surface.

The cliffs that typically enclose waterfalls and the caves that are often found behind them are the homes of many different creatures that otherwise may have little or nothing to do with falling water. These include a wide variety of insects, birds, bats and monkeys, many of which use the rocky ledges and cavities as refuges. Swifts are especially associated with waterfalls, where cliff ledges and cave walls provide ideal environments for these most aerial of birds.

While much of the plant and animal life in the vicinity of waterfalls has no direct relationship with these landforms, there are some that benefit from the microclimatic influence of the

A dipper in a
Swedish waterfall.

Petite Cascade, Madagascar, where the environment is ideal for plants that flourish in constantly moist conditions.

tumbling waters and the spray that they commonly generate. Reference has already been made to the splash zone in which some plants and animals flourish, but where the force of the fall is sufficient, the water is broken into very small droplets: atomized. The resultant spray is often carried aloft by air currents generated by the descending mass of water, or wafted or blasted away by the wind, depending on weather conditions. At many waterfalls, fine spray remains suspended in the air like a mist, descending slowly, but at great cataracts the spray may fall back to earth like heavy rain, beating down hard on the rocks and plants.

In this moist environment, many plants flourish, ferns, mosses and other greenery contributing greatly to the beauty of many waterfalls and their immediate environs. Where there is much spray, the vegetation in the vicinity of a large waterfall can be markedly different from that which normally occurs in the area. This is very apparent at Victoria Falls, where the spray rises as high as 300 metres in the air and falls as a constant shower on adjacent areas.

Nature's drama of life in and around waterfalls is an intriguing subject, but the main theme of this book is the relationship between human life and tumbling rivers and cascading streams. While the scientific study of waterfalls is one aspect of the human fascination with these landforms, for most waterfall lovers it is the aesthetic experience that gives the greater pleasure.

3 The Allure of Waterfalls

'Numberless falls of water, tumbling down the sides of steep
precipices, or rushing over the tops of huge stones in the beds
of the rivers, at once charm both the sight and hearing.'
Thomas Atwood, *The History of the Island of Dominica*[1]

The appeal of waterfall sights and sounds

In a television interview, Jane Goodall described chimpanzee
behaviour at an African waterfall: 'It's a very spiritual place.
Sometimes when the chimpanzees get there, you see their hair
become erect and they start like this – it's like a dance ... It seems
to me that this dance must be provoked by the same kind of
wonder and awe that we feel with these manifestations of the
wonder of nature.'[2] Perhaps the appeal of waterfalls that humans
experience is something we share with other primates: a trait that
may have evolved from a common ancestor millions of years ago,
and that some people, myself included, first experienced in child-
hood. In August 1838, recalling a holiday in Wales when he was
ten and a half years old, Charles Darwin wrote, 'I remember ... a
waterfall with a degree of pleasure, which must be connected with
the pleasure from scenery, though not directly recognized as such'.[3]

A search of the Web and a glance through holiday brochures,
guidebooks and travel literature show that waterfalls are widely
popular. Art galleries and illustrated books on the history of art
confirm that this liking is widespread among a variety of nations
and cultures past and present. It is a landscape taste that dates
back hundreds of years, and many writers have tried to explain
why we find waterfalls attractive. Their explanation is given
mainly in terms of aesthetics, largely related to the senses of sight
and sound; but, as relatively uncommon features of the natural
landscape, like caves, waterfalls have a curiosity appeal that other
beautiful objects encountered in everyday life do not possess. Many
common trees, for example, are widely regarded as beautiful but,

Eas Fors, Isle of Mull,
Scotland, forms a
textured arc as it shoots
over its rocky lip and
falls into Loch Tuath.

because we see them around us daily, we rarely take much notice of them.

An important theme in writings on the beauty of waterfalls is their ambiguous nature: falls are both transient and unchanging, the water flowing hurriedly by and descending rapidly while the rock face over which it tumbles in a continuous stream remains, unmoving. This is perfectly expressed by Guyanese writer Wilson Harris in his novel *Palace of the Peacock*: 'before them the highest waterfall they had ever seen moved and still stood upon the escarpment ... the immaculate bridal veil falling motionlessly from the river's tall brink'.[4]

For Rita Barton it is 'the movement and texture of falling water, which together create a loveliness of which one never tires. It is a beauty at once transient and reassuringly immutable. Moment by passing moment, season by season, the leaping river changes its voice, its mood, its aspect, in sensitive response to the varying elements around it.'[5] Edward Rashleigh wrote of 'the ever changing loveliness of waterfalls ... appealing in a thousand variations of sound to the ear, as their irised and constantly changing beauty appeals to the eye'.[6] Journalist Todd Lewan put it this way when describing a visit to the Cascata do Caracol in Brazil: 'Up above, the waterfall was doing what it always had: hurtling, plunging, tumbling, stretching, coming apart, regrouping, dropping, exploding, spraying, clouding. It never did it the same way, never had, never would.'[7]

We like waterfalls because they are pleasing to the senses, and, as eighteenth-century historian Thomas Atwood recognized, the senses that we most associate with the enjoyment of waterfalls are sight and hearing.[8] Sight is the dominant sense by which our species perceives the world around us. Waterfalls, with their constantly moving, bright, scintillating surfaces and textures contrasting with the dark rocks and foliage of their settings, inevitably attract the eye. What landscape architect John Motloch wrote about artificial cascades and waterfalls is equally true of those formed naturally in the wild. Their visual and auditory effects vary with 'the volume of water, the rate of flow, the condition of the edge over which the water flows, the height and nature of the fall, and the surface terminating the fall.'[9] The rate of flow affects the inertia of the water and the way in which it breaks when there is no underlying support. Where a powerful flow of water comes to an abrupt edge, it may shoot forward as it begins its fall, forming a graceful arc before dropping vertically into the pool or onto rocks below. When water drops over a smooth lip, it forms a sheet-like fall, one which may cling to the rock face if the current is slow, but when flowing rapidly over a rough edge it becomes turbulent and aerated. Depending on the volume and velocity of the water and the roughness of the edge over which it drops, the fall may have the appearance of a transparent sheet or a foaming blanket. This can change the colour and character of the rock face considerably, especially where the wet surface is absorbent. Irregularities at the edge of the fall may cause some parts of the rock face to remain dry, while protrusions into the flow can cause the water to splatter, the droplets sparkling in the light. When water falls freely, the greater its volume and the further it drops, the more intense is its impact as it strikes the surface below. Water falling onto a hard surface, such as a rock, generates a loud, harsh smacking noise for which Motloch appropriately uses the ono-matopoeic word 'splat'. Where it falls into a pool of water, the sound may have a more muted, deeper tone.[10]

Thus the volume, tone, pitch and reverberation of waterfall sounds depend on a number of variables, which include the quantity and velocity of water, the height of the fall, the degree to which it is interrupted by ledges of rock and the presence or absence of rocks or a pool at the foot, as well as its setting, for example, the way in which it is enclosed by cliffs and vegetation. Great waterfalls can generate a whole symphony of sounds, the total effect often being like powerful thunder. Depending on their size, form and location, waterfalls may, to borrow some other terms gleaned from the literature, 'roar', 'splash' or 'tinkle'.

Waterfall feelings, tastes, smells and colours

Sight and hearing are not the only senses by which we perceive waterfalls. Describing his visit to the Cascata do Caracol, Todd

Lewan wrote, 'Feeling cold breaths on my face, I slowed . . . My eyes closed. The mist gathered in droplets on my eyelids and ran down my cheeks'.[11] The writer not only saw the waterfall, he felt it, even at a distance. Draughts generated by the plunging water cooled his skin while on his face he felt droplets carried from the waterfall on currents of air. Beside a huge cataract we may even be able to feel the vibration caused by the impact of the great mass of water as it crashes down.

Shrouded in flying spray, we may sometimes taste a waterfall on our lips and there are those who claim that waterfalls can have their own distinctive smells. In damp environments, where spray keeps the surrounding rocks, soil and vegetation moist, conditions favour chemical and biological processes such as growth and decay, and odours are often particularly strong. In his description of High Force (see pages 109, 188), Victorian traveller Arthur Norway refers to the colours and sounds of the fall, but he also records, 'a sudden gust of wind sent a whiff of spray across my path . . . the indescribable fresh scent of falling water mingled with the odour of wet woods, and the keen air of the northern evening'.[12] According to Quest International, a company that explores the natural environment in search of fragrances that can be developed commercially, 'You get a characteristic odour around waterfalls. It smells like a mix of vegetation and minerals thrown off the rocks by the force of the water.'[13] A Quest natural products chemist who investigated the Grande Cascade in Madagascar claims that 'We have been able to identify some of the odorous ingredients . . . We found lots of familiar materials, mainly from the more water-soluble bits from trees and plants upstream, with resins, leaves, bark and moss growing around the falls.'[14]

What lies upstream of and beside a waterfall also influences its visual appearance in terms of colour. Waterfalls throughout the world are often stained brown or yellow, sometimes red, with the silt and mud that their rivers carry, particularly when in spate. Mud-stained waterfalls may not appeal to those who prefer crystal cascades to turbid torrents. Many British waterfalls have an amber tinge, derived from the peaty soils of the uplands from

A rainbow adds beauty to this view of the sublime Victoria Falls, here seen from Zimbabwe's side of the cataract.

which the rivers flow. Where rivers are fed by melting mountain glaciers, the water may have a milky appearance. This is caused by 'rock flour' formed by the grinding action of moving ice and carried in suspension by the glacial stream. The remarkable green colour of the water at Niagara Falls reflects the region's geology. Dissolved minerals and fine particles of rock are the main ingredients that give Niagara's water its distinctive colour. Some waterfalls, including those below the outflows of lakes which trap suspended river sediments, have transparently clear water, but this need not rob them of brilliant colour. Draining Lake Taupo, New Zealand's Waikato River becomes highly aerated as it seethes through a narrow rocky cleft before shooting over the Huka Falls in a curving mass of intensely blue water. An information sign there explains that this waterfall's unique colour 'is due to the very clear water reflecting blue light. The air bubbles in the water intensify the blue colour'.

As reflecting surfaces, waterfalls change colour with varying conditions of light and with changes in the immediate environment, such as seasonal variation in vegetation. Constant change is an essential part of the fascination of waterfalls.

Ever-changing waterfalls

Where geomorphological processes are particularly rapid, changes in the shape and position of a waterfall may be perceptible during a human lifetime, but it is change over much shorter time periods that contributes most to the appeal of waterfalls. Seasonal changes in flow and in the appearance of surrounding vegetation have already been noted, but waterfall scenes can alter dramatically within minutes, even seconds. When a brightly shining sun is suddenly hidden by passing clouds, the glaring brilliance of tumbling white water changes into a softer silver, the gentler light permitting our eyes to see the subtle tints in the water, while

Akron Falls, New York, in dry conditions.

Akron Falls in spate.

the bright colours of rainbows dancing in the spray vanish with the disappearing sunbeams. The emergence of a bright moon at night can create a lunar rainbow in the spray while illuminating the falls in a ghostly light. In the darkness of night it is the roar of the fall that most impinges on our senses, our ears now more sensitive to the subtle variations in the pulsating sound of the tumbling stream.

The volume of the water leaping over the falls can fluctuate rapidly in response to sudden changes in weather conditions upstream, as many venturing into or close to the stream bed have discovered to their cost. A heavy downpour over a watershed with rapid surface runoff may cause a surge in stream flow that utterly transforms a modest waterfall into a raging cataract. Changing volumes of water alter the appearance of waterfalls in several ways. Often the width of the lip varies with the volume of water in the channel. At Victoria Falls, the Eastern Cataract on the Zambian side largely ceases to flow in the dry season, while during the wet so much spray is generated by the swollen Zambezi as it tumbles into its chasm that the vast curtain of water is hidden. Falls that descend in one leap when in flood may cascade brokenly down the rock face when the flow is diminished. Falls that descend in a single sheet or column of water when in flood commonly split into two or more channels when the flow is less. Others become double or multiple falls as their rising waters spill over through new channels. India's Jog (or Gersoppa) Falls normally comprise four separately channelled waterfalls, each with its own name – Raja, Rani, Roarer and Rocket – but during the monsoon these merge into one enormous cataract hidden from view by dense spray. Another and more permanent change that must be noted here, as elsewhere in the world of waterfalls, is the large reduction of flow at Jog Falls in consequence of the construction of a dam upstream for power generation.

opposite: Waterfall in Milford Sound, New Zealand, on a windy day.

Jog Falls, also known as Gersoppa Falls, India, in flood. When the flow is reduced, this cataract divides into four separate channels.

Kinder Downfall in the Peak District moving in reverse in the teeth of a south-westerly gale.

Changes in water volume may alter the height of a waterfall as well as its width. Increased depth of water at the brink may add to the height but this is likely to be offset by rising levels in the river below. Indeed, these conditions can lead to a temporary reduction in height, possibly eliminating the falls entirely by submergence. This is most likely to occur where a waterfall drops into a narrow chasm in which the water level rises greatly because the flood cannot escape as quickly as it enters.

While changes in the volume of water can have astonishing affects on the appearance of waterfalls, there is something more playful about the way in which the wind influences the scene. A slight breeze is enough to set in motion the veils of spray that shimmer in front of many falls, while the column of falling water itself is often made to sway in a sinuous dance by the wind. More startling is the sight of a waterfall that flies upwards in the teeth of a gale. There are several waterfalls that are famous for the way in which, under suitable conditions, they are put into reverse as they leap from the brink. This is can occur when a strong wind blows directly towards the waterfall. Where the tumbling stream has eroded back into the cliff face, creating a natural funnel, the force of the wind is concentrated on the lip of the falls as at Kinder Downfall in the English Peak District. Here, when a southwest gale strikes the edge of the Kinder Scout plateau, the stream is sent skywards as a plume of spray. On the island of Mull, 'When the gales blow in from the Atlantic, some waterfalls . . . are checked in their descent and blown back in spray over the cliff tops by the wind, funnelling up the rock chimneys, until the whole headland appears to be smoking as if on fire.'[15] On the Hawaiian island of O'ahu, Waipuhui, meaning 'water blown up', is the name of another waterfall that is blown into reverse by the wind. In English this waterfall is named Upside Down Falls.

Where the temperature falls below freezing point, icicles often form around waterfalls, while spray-wetted surfaces on nearby objects like rocks and tree branches may become coated in ice. In severe winters, even the falls themselves may freeze into silent immobility. Small cascades turn to solid ice quite easily but it takes exceptional circumstances to completely freeze a

large waterfall. Large by modest English standards, Hardraw
Force, a 30-metre-high slender fall in the Yorkshire Dales, has
been known to turn into a hollow column of ice within which
water can be seen descending. On a far grander scale is the icy
spectacle of Niagara Falls in winter, when the cliffs are festooned
with gigantic icicles and the surfaces of nearby objects such as
trees and grass, railings and lamp-posts, glisten with a coating of
ice. At this time, ice floes plunge over both the Canadian and
American Falls, accumulating to form huge icebergs in the
frozen river below.

Apsley Falls, New South Wales, Australia, in spate.

4 The Beautiful, the Sublime and the Picturesque

'The cascades, the torrents, the rivers, and the rills, are enchantingly picturesque in their different features and exchange the sublimity or repose of their scenes, according to the variations of the seasons, or the turmoils of the elements . . . and in the rainy seasons these sublime and beautiful objects are very frequently to be met with.'
William Beckford, *A Descriptive Account of the Island of Jamaica*[1]

Beautiful or sublime?

In 1818, while exploring the New England area of Australia, John Oxley and his companions encountered two waterfalls on the Apsley river. In the words of his journal, they 'were lost in astonishment at the sight of this wonderful natural sublimity'. Continuing his description of the falls, Oxley, then Surveyor General of New South Wales, made the interesting observation that 'if the river had been full, so as to cover its entire bed, it would have been perhaps more awfully grand, but certainly not as beautiful'.[2] It is clear from Oxley's remarks that he considered the falls to possess aesthetic qualities both sublime ('awfully grand') and beautiful, noting that the beauty of the scene would be diminished with a much greater flow of water, while its grandeur and awfulness would probably be enhanced.

While most of today's visitors to waterfalls are probably unconcerned with aesthetic principles and theories, many of those in Oxley's day would have been familiar with the concepts of the Sublime and the Beautiful, some, no doubt, being acquainted with the considerable literature on the subject. In 1757 Edmund Burke published *A Philosophical Enquiry into the Origin of Our Ideas of the Sublime and Beautiful*, in which he attempted to make a clear distinction between these two aesthetic qualities. Comparing them, Burke wrote:

> For sublime objects are vast in their dimensions, beautiful ones comparatively small; beauty should be smooth, and

polished; the great, rugged and negligent; beauty should shun the right line, and when it deviates, it often makes a strong deviation; beauty should not be obscure; the great ought to be dark and gloomy; beauty should be light and delicate; the great ought to be solid, even massive. They are indeed ideas of a very different nature, one being founded on pain, the other on pleasure.[3]

opposite: Venezuela's Angel Falls (Kerepakupai Vená) is the highest waterfall in the world.

Burke's sense of the sublime was 'a sort of delightful horror, a sort of tranquillity tinged with terror, which, as it belongs to self preservation, is one of the strongest passions'. He recognized that 'the qualities of the sublime and beautiful are sometimes found united', but emphasized that we should distinguish the two, just as we always distinguish between black and white, however mixed or blended.[4] In the quotation above the use of the word 'great' in place of 'sublime' is worth noting but, as we shall see, in landscape aesthetics size is relative. Let us now consider the concepts of the sublime and the beautiful as they apply to waterfalls.

As abrupt changes in a river's course, waterfalls may be considered as inherently sublime, suddenness being one of the characteristics of sublimity, as gradual change is of beauty. However, a slight fall, because of its smallness, or a series of cascades, because of their more gradual descent, commonly achieve beauty. In the case of large falls, especially great cataracts such as Niagara, Victoria and Iguassu Falls, their astonishing magnitude arouses in the observer an overwhelming sense of the sublime, 'beautiful' being an inadequate word to describe them. The same may be said of very high waterfalls, such as Angel Falls, Yosemite Falls and Sutherland Falls, despite the relatively small volume of water that comes over the brink. Nevertheless, as William Wordsworth (1770–1850) observed, 'the sense of sublimity depends more upon form and relation of objects to each other than on actual magnitude'.[5] That said, if, as Burke asserted, beautiful objects are comparatively small, smaller waterfalls are more likely to possess the quality of beauty than are those of vast dimensions, which by definition are sublime.

Eighteenth- and nineteenth-century Romantics, however, discovered the sublime in landscape features, such as mountains and waterfalls, which are by no means among the world's giants. In Britain, for example, the modest Welsh, Scottish and Lakeland hills were described and portrayed in terms of the sublime. An important factor here is state of mind, particularly the observer's frame of reference based on past experience in similar situations. Hence, a waterfall may be perceived to be large – high, great in volume and so on – if the observer has seen nothing of its kind that is greater. The converse is also true.

A waterfall that does not impress the observer by its magnitude may charm the visitor with other aesthetic qualities. Near Rydal, in the English Lake District, there is a waterfall, small even by the modest standards of that area, which so delighted eighteenth-century visitors that Thomas West felt it worth quoting William Mason's description of it when writing his *Guide to the Lakes* (1778):

> Here Nature has performed every thing in little that she usually executes in her larger scale; and on that account, like the miniature painter, seems to have finished every part of it in a studied manner ... the little central current dashing down a cleft produc[ing] an effect of light and shadow beautiful beyond description ...[6]

While it is obviously the smallness of this modest cascade that puts it into the category of the beautiful rather than the sublime, qualities of light and shadow clearly contribute greatly to the beauty of the scene.

Darkness, gloom and obscurity are all terms associated with the sublime, while light and beauty go together. Waterfalls are typically found in dark, gloomy gorges, often largely hidden by umbrageous vegetation, but the white water reflects light very effectively and shafts of sunlight often illuminate the scene, their brilliance enhanced by the contrast with the surrounding shade. The moving water surfaces sparkle, and in the spray rainbows may form, their brilliant, scintillating colours merging into one another in a way that conforms perfectly with Burke's characterization of beauty.

The ruggedness and wild disorder of the setting contribute to the sublime quality of many falls where the descending water is broken on the irregular cliff-face, shivers on impact as it crashes onto tumbled rocks below and seethes madly in the plunge pool. In contrast, at the brink of many falls the water goes over in a smooth curve, the very epitome of beauty. While rugged rocks and tumbled boulders may abound, rock surfaces worn smooth and polished by the erosive action of the stream, some glistening wet, bring qualities of beauty to a sublime scene. Similarly, even at cataracts whose magnitude and rugged, gloomy setting make them utterly sublime, elements of beauty can always be found, including the delicate lacy texture of the falling water, the sparkling reflected light, and the rainbows that form graceful, colourful arcs in the spray. There is, too, a delicate beauty in the vegetation often found at waterfall sites. The foliage of overhanging trees and the ferns and mosses which flourish in the moist environment offset the sublimity of the massive, dark rocks and frowning cliffs that enclose the tumbling river.

The sublime and beautiful qualities of waterfalls are clearly identified in eighteenth- and nineteenth-century descriptions. Like William Beckford, the historian Edward Long was a Jamaican sugar planter, and both men wrote books about the island. The landscapes they described were seen through the eyes of Romantic connoisseurs. In his *History of Jamaica* (1774), Long describes the now despoiled falls on the White River near Ocho Rios. He begins by dwelling on the sublime aspect of the scene but towards the end he identifies other qualities that are compatible with the concept of beauty – softness, serenity, placidity and happiness – in contrast with the 'aweful' rage, gloom, fury and violence described in the earlier lines.[7]

The artist's eye: picturesque waterfalls

In his description, Long refers to 'the power of painting', and analyses the scene, 'the picture', as he puts it, in terms of a work of

Joseph Bartholomew Kidd, *The Windward Falls, Near Kingston, Jamaica*, 1830s, colour lithograph, engraved by W. Clark.

art.[8] In the art world, a picture of scenery is commonly described as a 'landscape', an English word derived from the Dutch *landschap*. This was originally a painting term, and is still used as such. In the early eighteenth century scenes in nature that were considered suitable subjects for pictures, or representations of them in art and literature, were termed 'picturesque'. Picturesque scenes presented the viewer with a well-composed picture, displaying suitable variety and harmony in form, colour and effects of light.

Sacrificed for electricity generation, Jamaica's Maggotty Falls occasionally regains its lost scenic beauty after heavy rain.

Later, the published works of British writers William Gilpin, Uvedale Price, Richard Payne Knight and others gave 'picturesque' a meaning that was less dependent on the artist's interpretation of nature, and more specific in terms of the qualities to which it referred. Characteristics of picturesque taste included a preference for rough textures, irregularity and the unexpected, all qualities associated with the sublime, but with less emphasis on the vast and awful. Rocky, broken terrain was preferred to smooth land surfaces, natural woodland and weatherbeaten trees to carefully maintained lawns and groves. Grottoes, ruins, rustic bridges, quaint cottages and mills were among the objects that appealed to those with a taste for the picturesque. So, at times, were figures in the landscape, such as peasants, hermits or exotic 'natives'; and waterfall scenes lend themselves to the inclusion of bathers and fishermen.

Another characteristic of the picturesque is partial concealment. Commonly hidden in narrow ravines, and approached by paths that wind over rugged, precipitous and sometimes densely wooded terrain, waterfalls are often partly screened from view by rocks and vegetation. Where the volume and height are sufficiently great, spray thrown up by the tumbling water may create a misty veil, concealing the waterfall further and contributing to the beauty, sublimity or picturesqueness of the scene.

Dorothy Wordsworth and her brother William had pronounced views on the aesthetics of waterfalls. Among the observations he made in his *Guide to the Lakes*, William noted, 'It is generally supposed that waterfalls are scarcely worth being looked at except after much rain, and that, the more swoln the stream, the more fortunate the spectator; but this, however, is true only of large cataracts with sublime accompaniments, and not even these without some drawbacks.'[9] Years earlier, Dorothy expressed similar sentiments in her journal of 1802 in which she recorded an account of a visit to Aysgarth Falls, Yorkshire: 'There was too much water in the river for the beauty of the falls.'[10] During normal periods, these falls on the River Ure are seen as an exquisite series of wide cascades in the form of a natural staircase, but after heavy rain they become a seething mass of water surging down a slope, their step-like formation lost beneath the swollen torrent. It is generally believed, however, that waterfalls are best seen when in spate, as guidebooks often claim.[11]

Modern landscape theories: arousal, prospect and refuge

Mental stimulation through the senses causes reactions that we may experience as pleasure or discomfort, even pain. Waterfalls can engender unpleasant as well as pleasant feelings. Some traditional people fear waterfalls that are believed to be the haunts of evil spirits. South Africa's Augrabies Falls is an example of this. Lord Curzon (1859–1925) described this remote cataract as 'the ugliest waterfall . . . in the world', a judgement he made without having seen the place for himself.[12]

In psychology, the term 'arousal' is used to denote the psychophysiological state of excitement or alertness in an organism. Fluctuations in arousal levels are reflected in changes in the electrical activity of the brain, muscular tension, the diameter of the pupil and in the circulatory and respiratory systems. Eminent psychologist Daniel Berlyne hypothesized that people prefer to experience a varying level of arousal, but one which varies in a range which avoids extremes of high arousal or overload, or low arousal or boredom.[13] This idea can be applied to waterfalls. Tumbling or cascading water, like waves breaking on a rugged coast, provide a higher level of stimulation than a placid lake but, despite their constant movement, they are in a sense unchanging. High arousal is stimulated by the kaleidoscopic patterns of falling water, the rising, swirling spray, the flash and sparkle of light, the pulsating noise; low arousal is occasioned by the stream's constant flow and descent and the water's unceasing sound: purling, splashing, roaring, thundering, according to the scale and form of the waterfall or the state of the weather. As a static composition, a waterfall in a setting of rocks and vegetation offers a spatial experience; the swift passage of water and the slow seasonal changes convey a sense of time. Thus part of the attraction of waterfalls may consist in this balance (emphasized by Berlyne) of immutability and change, stillness and movement, space and time.[14] As we have seen, this idea was intuited by lovers of natural landscapes long before psychologists got into the picture.

There are those who would suggest that the arousal which waterfalls can occasion may in part be of a sexual nature. 'What is the appeal of waterfalls?' asks travel writer E. King. 'Something sexual, perhaps? Are they a metaphor for passion, like waves thundering over jagged rocks, only more potent?'[15] There can be little doubt that waterfalls are associated with sexual passion and eroticism, something that has long been exploited by writers and artists.

Sexual symbolism in the landscape has been discussed by several scholars, including the geographer Jay Appleton in his seminal book *The Experience of Landscape* (1975). Appleton sought to explain landscape preference in terms of what he calls 'habitat theory'. 'Habitat theory postulates that aesthetic pleasure in landscape derives from the observer experiencing an environment favourable to the satisfaction of his biological needs.'[16] Though he said little about them, one can apply his ideas to the aesthetic analysis of waterfalls: clearly, fresh water is essential for life, and in the light of habitat theory signs of its presence can be expected to have a particularly strong appeal.[17] Indeed, there is abundant evidence to suggest that water bodies figure strongly in landscape preference. Waterfalls, therefore, may owe at least part of their landscape appeal to their being aqueous phenomena; and, unlike lakes and placid streams, they announce their presence not only to the eye, but also to the ear, indicating the presence of fresh water even when out of sight. They are part of the 'soundscape' as well as the visual landscape.[18]

More specifically, Appleton proposes 'prospect-refuge theory', central to which is the concept of 'hazard'. Water falling over cliffs and the rugged gorges in which waterfalls are commonly found may be perceived as hazards, as they are landforms and places that pose a threat to human safety. Appleton's prospect-refuge theory relates to our tendency to avoid danger, instead gaining advantage from our environmental conditions. This theory proposes that 'the ability to see without being seen is conducive to the exploitation of natural conditions favourable to biological survival and is therefore a source of pleasure'.[19] Much of Appleton's argument relates to hunting, which he, like many other scholars, believes to have been an important factor in the evolution of the human mind and behaviour.

It is hazard that bestows symbolic significance on refuge, and waterfalls may be recognized as an inanimate hazard, one which falls into three or four of the five sub-groups into which

Appleton categorized this type. Clearly, waterfalls belong to what Appleton called 'aquatic hazards', with their threat of drowning; and they are also a form of 'locomotion hazard', associated with movement, notably falling. Perhaps they may also be classed with avalanches and landslides as 'instability hazards'. Insofar as their flow is controlled by weather conditions, waterfalls are also related to the 'meteorological hazard' group, which includes rain, snow and ice.

Out of the five types suggested by Appleton, it is only 'fire hazards' that have no obvious direct connection with waterfalls. Even here, however, falling water and rising spray caught in bright sunlight and the play of reflected or refracted light on rock surfaces can be suggestive of flames. Writing in 1779, Goethe (1749–1832) described this phenomenon as observed at the Pissevache waterfall in Switzerland. He noted that as he climbed up towards the falls, he saw before him 'a constantly varying play of fire'.[20] Describing Aira Force in the Lake District, Mary Welsh wrote that the play of light in the spray made the cloud of water droplets 'flicker like flames'.[21] A different fiery effect is described in A. S. Byatt's novel *Possession*. At Thomason Foss near Whitby, 'what appeared to be flames of white light appeared to be striving and moving upwards' inside a small cave and on nearby boulders where the stream tumbled over.[22]

The swirling spray from waterfalls can also be suggestive of smoke from fires. In a letter written to his friend John Forster in 1841, Charles Dickens described the rain-swollen torrents which he saw tumbling down the precipitous sides of Glencoe, 'sending up in every direction spray like the smoke of great fires'.[23] For Arthur Conan Doyle, creator of Sherlock Holmes, the rising spray at the Reichenbach Falls, scene of the detective's apparent demise, resembled 'smoke from a burning house', and, as explorer David Livingstone recorded, Victoria Falls are known locally as 'Mosi oa Tunya', or 'The Smoke that Thunders'.[24]

Appleton identifies one more type of hazard, one that does not pose a direct threat to survival. This group he calls 'impediment hazards', including rivers, ravines and cliffs, features that are associated with waterfalls. These landforms are barriers to

Yosemite Falls, California, from John Gibson, *Great Waterfalls, Cataracts and Geysers* (1887). Framed by trees, this waterfall view exemplifies Appleton's prospect-refuge theory.

movement, possibly impeding escape from danger. Moreover, waterfalls have often impeded movement along rivers and valleys which otherwise might provide convenient routes.

Where a river flows from relatively open country and plunges into a secluded and typically, wooded gorge, the prospect-refuge symbolism can be very powerful. Conversely, a waterfall emerging from a hidden ravine and tumbling in full view down into a broad valley can evoke similar feelings. Commonly partly concealed and framed by rocks and trees, and seen from the sheltered viewpoint of a natural ledge on the side of a gorge or an observation platform, many waterfalls are quintessential manifestations of prospect-refuge symbolism. Geographer Jay Appleton has provided us with yet another possible explanation of our delight in waterfalls.

There are many who accept a totally different explanation of the positive influence that waterfalls appear to have on us – negative ions. Negative ions are invisible, odourless, tasteless molecules that are particularly abundant in the atmosphere at locations such as mountains, seashores and waterfalls. They are formed by the breakdown of atmospheric molecules caused by the sun's radiation and the movement of air and water. It is claimed that inhaling negative ions enhances one's feeling of well-being, generating the kind of elevated mood that many experience after a thunderstorm – or a bathroom shower. We are told that people who live near waterfalls live longer, healthier lives. In *The Ion Miracle*, Jean-Yves Côté explains that negative ions are electrons in the atmosphere which facilitate reactions between chemical components, thus helping our bodies to absorb oxygen when we breathe. He, like many others, makes some remarkable claims for the beneficial effects of negative ions, ranging from the alleviation or even cure, of asthma and allergies to the enhancement of sexual capacity.[25] Perhaps that is what puts a grin on the faces of people who visit waterfalls. In the following chapter we explore the association of waterfalls and sex further.

John Ruskin, *Cascade de la Folie, Chamonix,* 1849, watercolour sketch.

5 Waterfalls of Passion, Fountains of Love

'Waterfalls are moments of crisis: orgasm or heart attack.'
Deborah Tall, 'American Waterfalls'[1]

'Where erotic sweetness poured from the fountain basins.'
Simon Schama, *Landscape and Memory*[2]

Waterfalls of Venus

Describing a trip to Wales with her lover Dave, Joy, the protagonist in Nell Dunn's novel *Poor Cow* (1967), recalls, 'And we had it right on top of this waterfall.'[3] Later she writes in a letter to Dave, 'Oh God, I'll never forget the Water Fall in WALES it was terrific.'[4] Sexual activity at waterfalls appears to have a long history. According to classical tradition, it was at a waterfall, that of Afka on the Nahr Ibrahim, Lebanon, that Adonis met Venus. In Shakespeare's poem *Venus and Adonis* the goddess of love tempts the handsome youth with the words,

> Feed where thou wilt, on mountain or in dale:
> Graze on my lips, and if those hills be dry,
> Stray lower, where the pleasant fountains lie. (verse 39)

Elsewhere, in *The Passionate Pilgrim*, the poet refers more specifically to waterfalls as places where lovers meet:

> Live with me, and be my love,
> And we will all the pleasures prove
> That hills and valleys, dales and fields,
> And all the craggy mountains yields.
>
> There will we sit upon the rocks,
> And see the shepherds feed their flocks,
> By shallow rivers, by whose falls
> Melodious birds sing madrigals. (verse 20)

By Shakespeare's time the word 'fall' had the meaning of cascade or waterfall, but we probably should not imagine that the falls he had in mind in the scene he evokes here were more than modest rushes of water over low ledges of rock. One high though usually very slender waterfall that evokes a vision of love is Sri Lanka's Bambarakanda Falls, which cascades down a cliff-face into a deep, basin-like pool. Here the rock has been weathered and eroded to form a natural sculpture that is said to resemble two embracing lovers.

The beautiful waterfalls of some of the Pacific islands appear to have been associated with the erotic life when seen through the eyes of visiting Europeans in the eighteenth century. It was at one of these, Fautaua or Fataoua Falls, in 1871, that the French naval officer and romantic novelist Louis-Marie-Julien Viaud first saw Rarahu, the young Polynesian girl with whom he fell in love. Under the pseudonym Pierre Loti, he wrote the autobiographical novel *The Marriage of Loti* (1881), in which he describes how they met. Loti had taken a walk to the waterfall with its natural rock pool where 'all day there was company to be found; the beauties of Papeete', who chatted, sang and slept, sometimes plunging into the water, emerging to lie on the bank. Little wonder that, 'Thither seamen, on shore for a few hours, would come to take their pleasure'! While Loti was abandoning himself 'to this enervating existence . . . two little girls . . . two children . . . stole out to lie under the waterfall'. Apart from loin-cloths and head-dresses of leaves, 'their slender dusky bodies were otherwise bare'.[5] The prettier of the two friends was Rarahu. Above the waterfall was another pool, used by the two young girls as a 'private bath'. It was a 'rock-basin made on purpose, one might say, for the meetings of two or three intimate friends'.[6] While the reader is not actually told that Loti followed the girls there on that occasion, the implication seems clear. Martin Sutton comments in *Strangers in Paradise*, 'Soon, our lotus-eater hero is lost in the silken embrace of the fifteen-year-old Rarahu, and a new South Sea myth is born.'[7]

In the previous chapter, the sexual symbolism of waterfalls in the landscape was briefly noted, and *The Marriage of Loti* offers an appropriate example in the following extract:

We had reached the foot of the dark ravine, where the torrent of Fataoua falls with a sheer leap of more than nine hundred feet, like a great silver sheaf. In the depths of this gorge the scene was one of pure enchantment. The most lavish vegetation grew tangled in the shade, dripping and revelling in perpetual deluge; creepers clung to the steep, black walls, and among them grew tree-ferns, mosses and exquisite varieties of maiden-hair.[8]

Here, surely, is 'where the pleasant fountains lie'.

In complete contrast to Loti in his attitude to sex, English art critic John Ruskin (1819–1900) revealed his highly repressed sexual emotions in a remarkable pen and ink drawing he made of a waterfall, the Cascade de la Folie at Chamonix (see page 76). As biographer Wolfgang Kemp wrote, 'No psychoanalyst is likely to ignore this drawing, once he has read Ruskin's comments about his marriage, and about the anxiety he felt towards the supposed disfigurement of Effie's body.' Ruskin married Euphemia ('Effie') Grey in 1847, and the unconsummated union was annulled in 1854. Kemp describes the drawing thus:

> The massive curving slopes of the mountain are split by the cleft which the water has driven into their firm flanks. [Ruskin's book] *Modern Painters* refers to such clefts as scars, wounds, never-healing rifts. Or is it the dark unfolding of sex that we see? Sex as an open wound?[9]

Niagara of Passion

As Margaret Armstrong reveals in her biography of Fanny Kemble, the Victorian actress and writer had a 'mystical passion for waterfalls', one that was patently sexual in nature.[10] Fanny's response to these landscape features, as revealed in her own writings, is strongly suggestive of transports of sexual excitement:

> A frenzy of impatience seized upon me . . . I rushed down the foot-path cut in the rocks. Trelawny followed me. Down, down

The brink of Niagara Falls, viewed from the Canadian side.

I sprang, and saw through the boughs the white glimmer of a sea of foam – Trelawny shouted, 'Go on! Go on!' In another minute I stood upon Table Rock. Trelawny seized me by the arm and, without a word, dragged me to the brink. I saw Niagara – Oh God![11]

Ian Littlewood, author of *Sultry Climates: Travel and Sex*, is another writer who recognized this trait of landscape eroticism in Fanny Kemble's character. He, too, quotes an extract from her description of a visit to Niagara, which included a walk behind the falls:

> I feel half crazy whenever I think of it. I went three times under the sheet of water; once I had a guide as far a the entrance, and twice I went under entirely alone ... As I stood upon the brink of the abyss when I first saw it, the impulse to jump down seemed all but an irresistible necessity ... I think that it would be delightful to pass one's life by this wonderful creature's side, and quite pleasant to die and be buried in its bosom.[12]

The cataract's association with death as well as sex is clearly evident here, and it is not surprising that Niagara Falls and other waterfalls are often chosen by unhappy lovers as places to commit suicide. Japan's Kegon Falls is noted for this.

Niagara Falls has long been famous as a honeymoon resort, and this is discussed by Karen Dubinsky in her book *The Second Greatest Disappointment: Honeymooning and Tourism at Niagara Falls* (1999). Dubinsky writes:

> Niagara Falls undoubtedly did make visitor after visitor think of sex. The cultural depictions of Niagara as an icon of beauty, which were often expressed in terms of gender and heterosexual attraction, made way for forbidden pleasures of sexuality, romance and danger that countless travelers experienced while gazing at, or playing with, the waterfalls.[13]

It is interesting to note that 'visiting Niagara Falls' is one of the many jocular terms for female masturbation.

Niagara Falls are often depicted as a woman, sometimes as a mistress to whom the lover is ever tempted to return. As Dubinsky illustrates, descriptions of the falls abound in sexual imagery – the 'kiss' of the flying spray, the 'moan' of the rushing water, islands that rest on the 'bosom' of the waterfall while the water

Early romantic tourism at Niagara Falls, *c.* 1905–20.

itself 'caresses the shore', and 'writhes' and 'gyrates' as it surges on with 'mad desire' towards the 'passionate' whirlpool.[14] While the idea of waterfall as woman represents a male attitude, there is evidence that women rather than men associate waterfalls with sex and romantic love. A comparative study of the sexual fantasies of men and women found that among the most typical female responses was reference to romantic settings. These included forests, fields, beaches, islands and waterfalls. While fifteen per cent of the 409 women in the study mentioned romantic or exotic settings such as these, only four per cent of the 291 men did so.[15] For Deborah Tall, a waterfall can symbolize orgasm; for evangelical sexologists Linda Dillow and Lorraine Pintus, the

experience is 'like sliding down a mountain waterfall'.[16] It may be
no coincidence that female writers have featured prominently in
this discussion. Women are also well represented among authors
of waterfall guidebooks, one of which, *Romance of Waterfalls*, by
Barbara Bloom and Garry Cohen, includes information on 'kiss-
ing spots' at falls in the us Pacific Northwest.[17]

Many falls tumble into pools that are ideal for swimming and
diving, while the cascading water can provide a delightful shower
bath. Perhaps part of the reason for the erotic connotation of
waterfalls is their appeal as bathing places where inhibitions may
sometimes be stripped off together with the clothes removed in
preparation for the plunge. These are pleasures that can be enjoyed
alone with nature, or shared with a companion, or even as part
of a small crowd. Poolside rocks can be excellent places to sit or
lie in the sun and to gaze at the enchanting scene and, perhaps,
at other people there. The latter pleasure may be heightened by

A waterfall wedding
in Hawaii.

the scanty attire or nudity of the bathers. Indeed, some waterfalls
are favourite spots for 'skinny dippers', and are recommended as
such in guidebooks.

A 1990s Jamaica Tourist Board pamphlet, *Jamaica Honeymoon*,
contains a section headed 'Especially for Lovers' in which there
is the invitation to 'play together in the cascading falls that plunge
into seven freshwater pools' at one of the island's most popular
waterfall attractions, YS Falls (see page 207).[18] Indeed, waterfalls play a significant role in the wedding and honeymoon
sector of the tourist industry in the Caribbean and elsewhere.
Some offer 'Waterfall Wedding' packages for couples who want
to get married in waterfall settings.

Fountains of delight

Usually lacking the terrible power displayed by a large waterfall,
artificial cascades and fountains that ornament many parks,
gardens, courtyards and other designed open spaces are nevertheless often seen in terms of sexual symbolism and have long
been associated with erotic love.[19] This view has sometimes been
linked with visions of Paradise, even in the Christian tradition,
normally associated with a prudish attitude to sex. In their book
Heaven: A History, C. McDannell and B. Lang comment on an
engraving of *Paradise* by Hieronymus Cock (*c.* 1510–1570) after
a work by Hieronymus Bosch (*c.* 1450–1516). In this picture, a
group of 'saints' are depicted naked, playing in and around a
fountain.[20] Another group nude romp in a fountain is depicted
in *The Fountain of Youth* by Lucas Cranach the Elder, who is
well-known for his pictures of both religious scenes and seduc-
tive nudes. His paintings of *The Nymph of the Fountain*, of which
several versions are known, show a young woman wearing nothing
but a lascivious expression, reclining beside a fountain.

The Renaissance fountain, typically ornamented with sculp-
tures of nude figures, was an object with strongly erotic associations,
and this is no more clearly illustrated than in the book *Hypnerotomachia Poliphili* by Francesco Colonna and published in Venice in
1499. Originally written in highly stylized Italian and published

Lucas Cranach the Elder, *The Nymph of the Fountain*, 1518, oil on wood.

in French translation in 1546, a complete English edition was unavailable until the publication of Joscelyn Godwin's translation in 1999. As Godwin observes, 'There are few passages of the book that are not to some degree erotic, if we allow that Eros/Cupid/ Amor is the god who fires us with the desire for beauty of every kind.'[21] At every opportunity, Poliphilo, the narrator and protagonist of his erotic dream, 'indulges in an enumeration of detail that one might call fetishistic when he applies it to clothing or footwear, but is no less obsessive when the object is an elaborate fountain'.[22] Among the fountains described in great detail is one that features 'three nude Graces, made from fine gold . . . From the nipples of their breasts there flowed thin streams of water . . . the statues make a gesture of modesty, hiding with their left hands those parts that should be covered.'[23]

Another of Colonna's fountains is ornamented with a statue of a 'beautiful nymph . . . sleeping comfortably'. The fifteenth-century European preference for buxom women is well illustrated by the Venetian monk's reference to the fountain nymph's 'plump

The fountain of the sleeping nymph in a French edition of the *Hypnerotomachie, ou discours du songe de Poliphile* (1561).

thigh', and 'plump knees', and his appreciative comment that 'her thighs were suitably fleshy'. She also had 'narrow feet which tempted one to reach out one's hand to stroke and tickle them'. It was not only the beauty of the sculpted nymph that attracted attention but the way in which her body was used to effect in the design of the fountain.

> The nipples of her small breasts were like a virgin's, and from them spurted streams of water, cold from the right-hand one and hot from the left . . . The hot stream shot up high enough so as not to impede or hurt those who placed their lips to suck and drink from the right breast . . . and

the rest of her lovely body was enough to provoke even one made of stone, like herself . . . At her feet there stood a satyr, all aroused in prurient lust.[24]

The *Hypnerotomachia* is divided into two books, each with its own climax. Godwin explains, 'That of Book One takes place at the Fountain of Venus in the centre of the Island of Cytherea, where Poliphilo uses Cupid's arrow to tear the curtain that conceals the naked Goddess, and is then pierced by an arrow himself.'[25] The detailed description of the Venus Fountain extends over several pages, including the scene in which Poliphilo 'completed the penetration' with the arrow presented to him by a nymph:

No sooner had I taken the divine instrument than I was surrounded by a sourceless flame, and with urgent emotion I violently struck the little curtain. As it parted . . . I saw clearly the divine form of her venerable majesty as she issued from the springing fountain, the delicious source of every beauty. No sooner had the unexpected and divine sight met my eyes than both of us were filled with extreme sweetness, and invaded by the novel pleasure that we had desired daily for so long, so that we both remained as though in an ecstasy of divine awe.[26]

The association between fountains and the erotic life continued well after Colonna's time. Twentieth-century film-makers were among those who made use of fountains in this connection. In an early scene in Fritz Lang's 1927 silent science fiction classic *Metropolis* the hero Freder and a woman frolic amorously amid fountains in the Eternal Gardens of Pleasure. Decades later, the ornate baroque Trevi Fountain in Rome is the setting of voluptuous Anita Ekberg's famous midnight dip in Fellini's film *La Dolce Vita* (1960). When the Hollywood star's companion Marcello joins her in the fountain basin and moves to kiss her, the water ceases to fall. *La dolce vita*, the sweet life, is associated with paradise, and paradise – here on earth and hereafter – is what many of us seek, often unsuccessfully, as in Marcello's case. As we have seen

Anita Ekberg and
Marcello Mastroianni
in the Trevi Fountain,
film still from Federico
Fellini's *La Dolce Vita*
(1960).

in this chapter, fountains and waterfalls have long been included
in representations of Paradise: both the earthly kind, such as the
sensuous tropical islands that fired the European imagination, and
the heavenly one, as painted by Renaissance artists. The following
chapter explores further the association between waterfalls and
Paradise or Eden.

6 Paradise and the Hereafter

'They possess . . . that perfection of climate and most charming scenery which suggest . . . Paradise and the Garden of Eden . . . sparkling with numerous shining waterfalls.'
Henry M. Whitney, *The Hawaiian Guidebook for Travellers*[1]

Paradise gardens

Inspired by the Staubbach Falls in Switzerland, Goethe wrote his poem 'Spirit Song Over the Waters', which begins:

> The soul of man
> Resembleth water:
> From heaven it cometh,
> To heaven it soareth.
> And then again
> To earth descendeth,
> Changing ever.

Water is a recurring metaphor in representations of life, death and the afterlife. Depictions of Paradise, whether in some earthly Eden or in the next world, usually include images of water, often in the form of tumbling streams.

In cultures around the world, the traditional concept of Paradise, the blessed world of the afterlife, is a garden; one that is well-watered with rivers, the verdant landscape commonly enlivened with waterfalls and fountains. Ronald King conceived the history of the world's gardens as *The Quest for Paradise*, the title of his book published in 1979. This work traces garden history from the great civilizations of the ancient world to the twentieth century. In King's account, waterfalls, cascades and fountains play an important role as essential features of artificial earthly paradise gardens.

The concept of paradise as a garden is shared by the Christian and Islamic faiths. The biblical Garden of Eden, watered by a river that divides into four streams, became Paradise Lost when Adam and Eve were banished for their sin. In the Quran, 'garden' or 'gardens' are the words usually used to denote paradise, and the afterlife of the righteous is frequently depicted as being enjoyed in gardens through which rivers flow. In addition, several verses of the Quran make specific reference to fountains in the gardens of paradise. In Christian Europe, the Garden of Eden was conceived as containing the Fountain of Life, often represented in illuminated manuscripts as an ornamental structure. Strange fountains feature prominently in two of the panels of Hieronymus Bosch's triptych *The Garden of Earthly Delights*, in the Earthly Paradise, or Garden of Eden, and in the Garden of Earthly Delights.

Milton's description of the landscape of Eden in *Paradise Lost* (1667) may owe something to this painting. The wild landscape he evokes includes a fountain and 'many a rill' which 'Watered the Garden; thence united fell / Down the steep glade . . .'. Elsewhere, Milton tells us that Eden is a place where 'murmuring waters fall / Down the slope hills'.[2] This is not to suggest that waterfalls or fountains play an essential role in the concept of paradise, but in imagined images of Eden and in designed landscapes that are attempts to recreate paradise on earth, waterfalls and fountains are very commonly found. With its forested mountains, caverns, fast-flowing river and tumbling streams, Milton's earthly paradise displays characteristics associated with the sublime and the picturesque, the kind of landscape which became fashionably popular in the century following the publication of *Paradise Lost*. It should be remembered that in Milton's day it was widely believed that mountains were deformities of nature, the world having been originally created perfectly level and smooth. As Classicism and then Romanticism increasingly influenced landscape taste, it was, perhaps, Arcadia or Elysium rather than the biblical Eden that designers sought to create.

In the nineteenth century much of the European and American population lived in what Lewis Mumford described as a 'Paleotechnic Inferno'.[3] Suffering daily among the 'dark Satanic mills', described by William Blake *c*. 1808, of squalid towns and cities, many people looked to the wilderness for an escape

John Martin, *The Plains of Heaven*, c. 1851, oil on canvas.

to paradise. When the wilderness itself was threatened by the vile hand of man, it often became enclosed as a park. Later, national parks began to be established. In *Wilderness and Paradise in Christian Thought* (1962) George Williams observes that 'In this development the wilderness, from being a desert of death and devils, has, with its variegated flora and fauna living in peace, been converted into a kind of paradise.'[4] An important factor in the origin of the American national parks movement was the public reaction to the commercialization and spoliation of Niagara Falls. Waterfalls were among the most significant features of the natural landscapes protected in many state and national parks of the USA and elsewhere. Notable early examples include the magnificent falls of the Yosemite Valley, put under public protection as a state park in 1864 and included in the Yosemite National Park in 1890, and those of Yellowstone, generally recognized as the world's first national park. The latter was established in 1872 for the preservation of a landscape that includes spectacular waterfalls, as well as those fountains of nature, geysers and hot springs. Among those who supported the establishment

of national parks in the USA were the railroad companies, which anticipated a boost to their business as more people were encouraged to see for themselves the natural wonders that were being preserved for public enjoyment.

Thomas Cole, *Distant View of Niagara Falls*, 1830, oil on wood. Cole's painting depicts the cataract in its unspoiled natural state, before the impact of industry and tourism.

Tourist paradise

Escape to paradise on earth has long been a motive for tourism, and tourist industry promotion commonly involves the evocation of paradisial places. The tourist paradise is commonly a tropical island, typically with palm-fringed sandy beaches lapped by the blue sea. Very often, the tourist image of paradise island includes a verdant mountain landscape enlivened by cascading streams and idyllic waterfalls. A guidebook of 1875 praises the Hawaiian Islands' perfect climate and picturesque scenery, including tropical forested mountains and 'numerous shining waterfalls', which together evoke 'Paradise and the Garden of Eden'.[5]

David Lodge's satirical novel *Paradise News* makes much of this now hackneyed paradisiac image of Hawaii with its 'swimming, surfing and picnicking all year round ... wonderful volcanoes, waterfalls, rainforests' and other earthly delights.[6] One character, a British anthropologist whose research field is tourism, is gathering material for his next book, on 'tourism and the myth of paradise'.[7] In addition to collecting tourist brochures with the paradise theme, he makes lists of the names of the many Hawaiian businesses he encounters that include the word 'Paradise'. Another character in the novel describes a day trip to 'The Garden Isle' of Kauai where he and his party 'kept getting glimpses of beautiful beaches ... but we were never allowed to get out of the minibus and explore them because we were always tearing off to another bloody waterfall.'[8]

Hawaii's numerous waterfalls are part of the tropical island paradise image promoted by the tourism industry. Kaluahine Falls, seen here, cascade straight into the sea.

Under the heading 'PARADISE welcomes you' one journalist in a recent article typically describes the attractions of the South Pacific islands as including 'coral reefs, lush mountains, waterfalls and spectacular islands and atolls'.⁹ Tropical paradise vacation lands are not confined to islands, however. In his article 'Selling Paradise and Adventure', geographer Gordon Waitt observes, 'for those in search of a romantic paradise, the Australian landscape is represented with the symbolic qualities of a Garden of Eden. In the visual text, paradise is symbolised by vibrant colours, sunshine and sunsets, exotic animal and bird-life . . . waterfalls . . . tropical beaches and palm trees'. Waitt's reference to 'a romantic paradise' appears to be less related to the Romanticism associated with picturesque landscape tastes than to idealized love, for he goes on to mention 'tourist couples . . . hand-in-hand within these settings' and the promised 'awakening of unknown heterosexual romance'.¹⁰

The hereafter

Waterfall-lovers should be pleased to learn that cascading rivers and streams may be among the delights that await them in the next life. As previously discussed, in Western cultures, waterfalls are a recurring feature in representations of the blissful place to which good souls go after death. Among the Christian sects that proselytize using images of a paradise world after death are Jehovah's Witnesses, who operate under the official name The Watchtower Bible and Tract Society. According to the numerous publications of this organization, 144,000 human beings will be chosen to go to heaven, but millions more righteous people 'will possess the earth, and they will reside forever on it' (Psalm 37:29). In an illustrated tract titled *The Government That Will Bring Paradise*, heaven, 'God's Kingdom', is represented by what looks like a mystical walled city with towers. The earthly paradise, to which most of the righteous can look forward, is depicted in pictures showing idyllic landscapes evocative of national parks, with forested mountains, tumbling streams and colourful flowers. Three of the illustrations feature waterfalls in the scenic background.¹¹

I do not suggest that the Jehovah's Witnesses or any other Christian believers necessarily expect to see waterfalls in the next life. Probably, this reflects contemporary landscape tastes among the kinds of people from whom the Jehovah's Witnesses are recruited. One more example of waterfalls in the popular image of paradise may be familiar to devotees of the television cartoon series *The Simpsons*. In one episode, 'Simpsons Bible Stories', a dream sequence shows the characters Homer and Marge in the roles of Adam and Eve. The Garden of Eden in which they find themselves also has a waterfall, one down which Homer, as Adam, dives in order to impress his mate.[12]

In the afterlife, waterfalls are not only found in heaven. While it may not surprise us when Milton tells that 'Hell . . . spout[s] her Cataracts of Fire', the waterfall that Dante (1265–1321) encounters in Hell, as described in his *Inferno*, is something perhaps less expected. Here it is falling water, not fire, that descends to the eighth circle of Hell, in which spirits are doomed to dwell 'beneath the fierce, tormenting stream'.[13] The inspiration for this infernal waterfall was the Cascata dell'Acquacheta, which is close to the monastery of San Benedetto in Alpe where the Florentine poet spent part of his exile. No doubt, the ceaseless wild tumult and angry roar of the Apennine waterfall impressed Dante as symbolic of an abyss of eternal punishment and torment. This is certainly how another Italian waterfall, the Cascata delle Marmore at Terni, impressed the Romantic poet Byron, who described it as

> The hell of waters! where they howl and hiss,
> And boil in endless torture; while the sweat
> Of their great agony, wrung out from this
> Their Phlegethon, curls round the rocks of jet
> That gird the gulf around, in pitiless horror set.[14]

Roben Waterfall at Oyama, Sagumi province, Japan. Woodblock print by Katsushika
Hokusai, 1832–3.

7 Waterfalls and the Creative Mind: Literature and Art

'A thousand painters and poets all down the centuries have caught their message and have been inspired by it.'
Edward Rashleigh, *Among the Waterfalls of the World*[1]

Myths and legends

Waterfalls have inspired the imagination since they were first beheld. We encounter them in myth and legend, literature and painting, music, photography and film. I have already discussed how our present experience of waterfalls and the way we imagine them in the hereafter both draw extensively on the arts. This chapter is the first of two that present a more detailed consideration of waterfalls as they are represented by the creative mind.

In the heart of Africa, original home of humankind, the Victoria Falls region has been occupied by humans and their hominid forebears since at least the arrival of *Homo habilis*. It is their primitive stone tools that mark the beginning of the Palaeolithic or Old Stone Age. On the Zambian side of Victoria Falls there is a museum, built over an archaeological site, where specimens of prehistoric stone tools, abundant in this area, are displayed. Some date back over two million years to roughly the time of the birth of Victoria Falls, when the Zambezi first spilled over the edge of the basalt plateau into which the river has since cut its zigzag gorge. It is evident, therefore, that the cataract's two million-year history, during which the falls occupied eight different sites, was witnessed through the eyes of genus *Homo*. The ever-present sight and sound of the great cataract inevitably influenced the thoughts of the local inhabitants. Here, as elsewhere in the world, there was a belief in a divinity who dwelt there. To conciliate this malevolent being, the local people would make offerings, casting a bead necklace, bracelet or some other sacrificial object into the abyss.

Among the earliest arts is that of storytelling, which, like our enjoyment of landscape, probably had its origins in the distant past when our ancestors were hunters and gatherers. With the development of speech, it is possible that word pictures were images far more common than those painted on cave walls or carved on bone or stone surfaces. All over the world, traditional tales of adventure and of the supernatural survive from the past, many of them associated with places and natural landscape features regarded with awe by the local people. Where waterfalls occur, they are commonly associated with myths and legends of this kind.

Sometimes the falls themselves have mythical origins, as in the case of the Angel Falls (see page 65), which the Pémon Indians of Venezuela believed was created from the tears of an evil god defeated in battle by his good rival. In many parts of the world it is traditionally believed that waterfalls are the homes of gods, spirits, sprites and the like. The Jívaro or Shuar Indians of Ecuador, a headhunting people made famous in *The Jívaro: People of the Sacred Waterfall* by Michael Harner (1972), believed waterfalls to be the gathering places for the souls of ancestors, their presence indicated by the movement of spray wafted about by the breeze.[2] The Naiads of Greek mythology, the trolls and demons of Scandinavian folklore, the Japanese spirits known as Kamis and the Mexican Chaneques, old dwarfs with children's faces, all frequent waterfalls. In Africa, South America and Australia snake monsters lurk in waterfall pools, while Niagara Falls is the dwelling place of the Great Spirit whose habitation is the spray and whose voice can be heard in the roar of the tumbling waters. The ancient Egyptian female deity Satet was goddess of the Nile Cataracts, as well as of the hunt, the inundation and fertility, a combination that is perhaps suggestive of the sexual symbolism of waterfalls.

In myth and legend, waterfalls commonly have associations with lovers. One story relates to Sri Lanka's Diyaluma Falls where, in her attempt to escape from home to join her exiled lover, a noble lady fell to her death while trying to climb a cliff. Moved to pity by the tragedy, the gods caused a stream to gush from the mountainside, creating a waterfall that veiled the sad scene. A very different tale is told of Jamaica's Llandovery Falls. Here a Spanish soldier, in love with a Taino chief's daughter, is said to have met his watery grave through the cunning of the beautiful girl who rejected the advances of the loathed invader. Folk tales about waterfalls are not always as romantic as these. Guyana's Kaieteur Falls (see page 40) is translated as 'Old Man's Fall', the name derived from the traditional local story about an old man who had become a nuisance to his tribe. To rid themselves of him, they set him adrift in a flimsy craft above the falls, over which he plunged to his death. A more heroic version of the story has an old chieftain responding to a sign in the form of a flight of birds and sacrificing himself to the Great Spirit, Makonaima, by canoeing over the cataract, a selfless act performed for the good of his tribe.

Sacrifice, as well as suicide and murder, are acts long associated with waterfalls, and the stories range from the tossing of images of the old gods into Godafoss, Iceland, when Christianity came to that country, to the ritual, attributed to the Native Americans living in the Niagara region, of sending a beautiful maiden in a canoe over the famous Falls. It is by the pool at the foot of the Afka Falls, in modern Lebanon, that Adonis is said to have been killed by a boar sent by jealous Mars, and was buried there close to where he met Venus. The writing-down of legends such as these was the beginning of classical literature, preserving for us stories that would otherwise have been lost. In other cultures, too, the evolution or introduction of the written and, later, the printed word gave rise to diverse literatures, including poetry, prose and drama.

The Far East

Probably nowhere is the influence of landscape in the arts stronger than in China. In a country where *shanshui* (literally mountain and water) painting flourished over a thousand years ago, and, from even earlier, poetry and prose abounded in landscape imagery, waterfalls have long been favoured landforms.

Fan K'uan (1254–13
*Travellers Among
Mountains and Streams,*
ink and colour on silk.

From ancient times, Chinese scholars and artists have travelled far to stand or sit in contemplation before these wonders of nature. The Tang period (AD 618–907) is considered the golden age of Chinese poetry, and among the many poems that survive from that time landscape is a constant theme. Mountains, forests, lakes and tumbling streams provide settings for action, evoke and reflect moods or are used as metaphors. Waterfalls often feature in this rich literature. In *A Message to Commissioner Li at Zizhou*, Wang Wei tells of 'a night of mountain rain' after which 'hundreds of silken cascades' appear, while Li Bai's *Hard roads in Shu* dramatically describes how 'a thousand plunging cataracts outroar one another / And send through ten thousand valleys a thunder of spinning stones'.

In Chinese painting, as in its poetry, waterfalls are often prominent. Laurence Binyon, an eminent Western authority on Far Eastern art, wrote, 'It is in landscape and the themes allied to landscape, that the art of the East is superior to our own.'[3] Centuries before the flowering of European landscape painting in the Renaissance, Chinese and Japanese artists achieved a mastery of the genre that was, perhaps, never exceeded in the West. Waterfalls were often included in the mountain and river scenes favoured by the painters, and were sometimes the main subject of their work. Several of the great masterpieces of Chinese painting portray waterfalls. Examples include the famous *Travellers Amid Streams and Mountains*, attributed to Fan K'uan,

Ma Yuan
(*fl. c.* 1190–1225), *Scholar Contemplating a Waterfall*, ink and colour on silk (album leaf).

who lived about a thousand years ago, and *Early Spring*, dated 1072, by Kuo Hsi (Guo Xi). A painting which conveys something of the attraction that waterfalls had for the cultured Chinese elite is *Scholar Contemplating a Waterfall*, by Chung Li (active *c.* 1480–1500). Another is Wen Zhengming's (1470–1559) *Aristocratic Scholars in a Lonely Ravine*. In this picture, an elegantly robed gentleman sits on a rock overhanging a mountain stream, apparently listening to the sound of the waterfall opposite. In some paintings, waterfalls dominate the scene, as in Wen Zhengming's *Pine-trees and Cypresses by a Waterfall*, and Mei Qing's (1623–1697) *Waterfall Plunging from an Overhanging Cliff.*

Very much in the Chinese tradition is the Japanese painting *Landscape* by Shabun (*c.* 1390–1464). This hanging scroll, with its mountains, cliffs, pavilions and graceful waterfall, reflects the artist's familiarity with Chinese art. One of Japan's most famous paintings is of a waterfall, *The Waterfall of Nachi*, attributed to Kanaoka (Kamakura Period, 1185–1333). This is a sacred site, dwelling of a local Shinto deity. Strikingly different in its pronounced horizontality as well as its pictorial style is Maruyama Oko's (1733–1795) screen painting of *Hozu River*, with its broken falls and rapids. By this time, landscape painting had become an important genre in Europe, and much European literature, too, was taking inspiration from nature and the landscape.

In other parts of Asia waterfalls may not have played the important role in art that they did in China and Japan, but some Indian examples are of special interest because of their representation of landscapes in carved stone. Dating from about 100–200 BC, the rock-cut Rani Gumpha monastery in Orissa is decorated with reliefs, some of which depict landscapes with waterfalls. Carved in stone bas-relief, *yakshinis* – female nature spirits – are commonly found in Indian temples, sometimes represented dancing, plucking blossoms, or bathing under a waterfall.

Europe

At least as far back as the time of classical Greece and Rome, landscape has played a role in European art. This was expressed in painting and poetry as well as in architecture; asscociated with the latter was the development of landscape design. Mountain and river scenery feature prominently in the *Idylls* of Theocritus (*c.* 310–250 BC), in the first of which there is a scene by a waterfall that makes sounds evocative of music as it tumbles over rocks. The heritage of classical culture influenced all aspects of European Renaissance art, including literature and painting. The landscapes evoked by Shakespeare, Milton and others echo those of the classical writers and, like the ones described in ancient texts, are typified by woods and mountains watered by springs, fountains and cascading streams. Unlike Dante, many of the writers and painters who included waterfalls among their subjects may never have seen a real waterfall in their lifetime, depending for their inspiration on the works of others who were familiar with them.

Following in the landscape painting tradition developed by artists like Joachim Patinir (*c.* 1480–1524), Albrecht Altdorfer (*c.* 1480–1538) and Pieter Bruegel the Elder (*c.* 1525–1569), the Flemish painter, Kerstiaen de Keuninck (*c.* 1560–1632) may be regarded as a European pioneer in the pictorial representation of natural waterfalls. His painting *A Mountainous Landscape with a Waterfall* probably dates from about 1600. While the left side of the picture is occupied by sharp, rocky peaks, the right is dominated by a steep, wooded hillside, down which tumbles a fine cascade. The composition is quite similar to that of the same artist's *Landscape with Acteon and Diana*, in which, instead of a waterfall, there is an elaborate fountain decorated with classical sculpture, at the foot of which is a group of nude bathers.

In the seventeenth century, waterfalls became increasingly common in landscape paintings, often as details in mountain scenes and sometimes as the principal subject. This motif was popularized by painter and engraver Allaert van Everdingen (1621–1675), who travelled to Norway and Sweden, returning to Holland in 1644. His pictures of mountainous landscapes with torrents and waterfalls inspired Jacob van Ruisdael (*c.* 1628–1682), who never visited Scandinavia, but, from the late 1650s onwards, painted many waterfalls, some of which are described

Jacob van Ruisdael,
*Landscape with a
Waterfall*, c. 1670,
oil on canvas.

as 'Norwegian'. The word 'waterfall' appears in the titles of at least sixteen of Ruisdael's paintings. Waterfalls appear in other landscapes of his, too. The artist's fascination with falling water is also revealed in his paintings of watermills and open sluices, the latter of which are, in effect, artificial waterfalls. It is tempting to suggest that the apparent strong preoccupation with falling water among artists of the Low Countries may be related to the flatness of their native land and the consequent absence of waterfalls, something that could have added to their particular fascination.

As evidenced by their representation in European literature and painting, the popularity of waterfalls appears to have grown considerably between the seventeenth and nineteenth centuries. This was the period of the Grand Tour. At first, the wealthy young travellers went, often in the charge of personal tutors, mainly to gain knowledge and experience of the manners and cultures of foreign countries. They usually visited places and saw sights of

cultural significance, great cities and famous buildings, classical ruins and important art collections, for example. As the taste for landscape developed, more tourists went to areas of scenic beauty. With the rise of Romanticism, those in search of the sublime travelled to regions where wild, rugged scenery was to be found, notably in the Alps. Here, in addition to lofty snow-capped peaks and rocky precipices, there were cascading torrents, many with impressive waterfalls. When war discouraged travel on the Continent, British travellers turned increasingly to the scenic beauties of their own country, leading to the development of tourism in areas like the Lake District, North Wales and the Scottish Highlands, all of which boast picturesque, sometimes sublime, waterfalls among their landscape attractions.

Artists inspired by their scenic beauty sought to capture waterfalls in pencil, ink and paint. Public demand provided a ready market for the drawings and paintings; prints of these scenes were produced in large quantities and sold to the many people who could not afford to buy the originals. Many prints of

Meindert Hobbema, *Landscape with a Watermill*, *c.* 1666, oil on wood.

Joseph Anton Koch,
*Waterfall in the
Bernese Alps*, 1796,
oil on canvas.

scenic views, including waterfalls, appeared in the travel books and guides that were being published to meet the increasing demand of tourists, as well as armchair travellers. Francis Towne (1739–1816) was one of the English landscape painters whose work encompassed scenes both in Britain and on the Continent. They include views of the Lake District and the Alps. Working mainly in pen and ink and watercolour, Towne travelled widely, his pictures including several waterfall scenes. A waterfall can be seen beside the road in his watercolour *Going up Mount Splügen* (1781), which shows part of a popular route taken by tourists

Francis Towne,
The Falls of Terni,
1799, watercolour.

crossing the high Alps between Switzerland and Italy. Other
scenes painted by Towne include waterfalls in Devonshire and
the Lake District.

The ability of artists to capture waterfalls on paper and canvas
was something to which the Victorian art critic John Ruskin gave
considerable attention. Of a waterfall study by the English
painter and engraver Thomas Girtin (1775–1802), Ruskin wrote,
'every sparkle, ripple and current is left in frank light by the steady
pencil which is at the same instant, and with the same truth,
drawing the forms of the dark congeries of channelled rocks,
while around them it dispersed the glitter of their spray.'[4] In his
book *Modern Painters*, Ruskin devoted a dozen or so pages to what
he called 'rock-drawing' and 'torrent-drawing'. This section is

J.M.W. Turner, *High Force or Fall of Tees*, *c.* 1816–17, watercolour.

largely concerned with the representation of waterfalls, emphasizing the artist's skill in rendering the tumbling water and the eroded rocks. For Ruskin, no one exceeded Turner as a painter of waterfalls. 'Turner was the only painter who had ever represented the surface of calm or the *force* of agitated water. He obtains this expression of force in falling or running water by fearless and full rendering of its forms.'[5]

Ruskin goes on to analyse Turner's rendering of High Force, in which, though the rock basin is obscured by rising spray, 'the attention of the spectator is chiefly directed to the concentric zones and delicate curves of the falling water itself; and it is impossible to express with what exquisite accuracy these are given'.[6] Turner succeeded in capturing the characteristic form of a powerful stream descending without impediment from a narrow channel, an achievement that Ruskin felt would be difficult to find an equal to in the world of art.

Where water takes its first leap from the top, it is cool,
and collected, and uninteresting, and mathematical; but it
is when it finds that it has got into a scrape, and has farther
to go than it thought, that its character comes out: it is then
that it begins to writhe, and twist, and sweep out, zone after
zone, in wilder stretching as it falls, and to send down the
rocket-like, lance-pointed, whizzing shafts at its sides,
sounding for the bottom. And it is this prostration, this
hopeless abandonment of its ponderous power to the
air, which is always peculiarly expressed by Turner.[7]

It is interesting to compare the Victorian art critic's words with
those of the twenty-first century travel journalist, Todd Lewan,
quoted in chapter Two.

High Force in the North Pennines is one of the many waterfalls in northern England that Turner drew on the spot and later
painted, his pictures being reproduced in large quantities as
engravings. Turner also painted falls he saw on his Continental
tours, including the Rhine Falls and Reichenbach Falls, both
in Switzerland.

Like Turner, the poet William Wordsworth, was familiar with
many waterfalls in both Britain and on the Continent. Among
those that he saw on his European tour of 1820 was the Swiss
fall which inspired the poem 'On approaching the Staub-bach,
Lauterbrunnen' – 'This bold, this bright, this sky-born WATERFALL!'
Waterfalls also feature in several other poems by Wordsworth,
including 'The Waterfall and the Eglantine' and 'The Idiot Boy'.
The Lake District waterfalls of Lodore and Aira Force are mentioned by name in 'An Evening Walk Addressed to a Young Lady'
and 'The Somnambulist', respectively.

Perhaps the best known of all waterfall poems is that of
another of the Lake Poets, Robert Southey (1774–1843), whose
onomatopoeic 'The Cataract of Lodore' was written 'for the nurs-
ery' in 1820. It long remained popular, and has been published
as illustrated books for children. The poem starts with a description of the infant stream which gradually grows into a torrent:

> The cataract strong
> Then plunges along . . .

later

> Rising and leaping,
> Sinking and creeping,
> Swelling and sweeping . . .

continuing in this vein until finally,

> All at once and all o'er, with a mighty uproar,
> And this way the water comes down at Lodore.

From this lively poetic evocation of the celebrated cascade it is evident that Southey (who lived in the Lake District) had studied these falls when they were in spate because, for much of the time, the stream flow there can be disappointingly slight.

For the eighteenth-century poet Mary Robinson, however, it was not the sight and sound of the great cataract in Switzerland that inspired her write about the Rhine Falls, but de Loutherbourg's picture of it. In 'Lines inscribed to P. de Loutherbourg, Esq. RA On seeing his Views in Switzerland, &c. &c.', Mrs Robinson expresses her emotions aroused by an image on canvas which represents a real place:

> Where the fierce cataract swelling o'er its bound,
> Bursts from its source, and dares the depth profound.
> On ev'ry side the headlong currents flow,
> Scatt'ring their foam like silv'ry sands below.

Typical of the sublime imagery of the period is Robinson's shared usage of descriptive phrases like 'awful grandeur', 'craggy height', 'torrents roar', 'Tremendous spot!' and, of course, 'views sublime'.

For some poets it was not any particular fall that was the subject of their verse, but instead they used the concept of the waterfall symbolically. A contemporary of John Milton, Henry Vaughan (1622–1695), wrote a poem titled 'The Waterfall' that made no reference to any particular place and, indeed, contained little landscape description, instead using 'this loud brook's incessant fall' as a metaphor in this spiritual work about life, death and existence beyond the grave.

Not only in English literature were waterfalls used as a metaphor for human life and death. This is also found, for instance, in the work of Goethe, the leading figure of the German Romantic movement. In the first scene of Part Two of Goethe's most famous dramatic poem, Faust awakes to find himself lying in a meadow by a waterfall:

> And so I turn, the sun upon my shoulders.
> To watch the waterfall, with heart elate,
> The cataract pouring, crashing from the boulders,
> Split and rejoined a thousand times in spate;
> The thunderous water seethes in fleecy spume,
> Lifted on high in many a flying plume,
> Above the spray-drenched air. And then how splendid
> To see the rainbow rising from this rage,
> Now clear, now dimmed, in cool sweet vapour blended.
> So strive the figures on our mortal stage.
> This ponder well, the mystery closer seeing;
> In mirrored hues we have our life and being.[8]

Goethe often used the waterfall as a symbol of human life, and in Faust's vision the falling water symbolizes continuity and change, unity and variety. Earthbound though the water was, the spray rose heavenward.

America

The eighteenth and nineteenth centuries were a period in which European influence spread rapidly round the globe. While the colonization of America advanced and the United States won independence from Britain, European settlement of Australia and

New Zealand began. There, and in other parts of the world, artists and writers recorded the strange new landscapes they experienced, their earlier works being largely images as seen through European eyes, influenced by contemporary concepts of the beautiful, the sublime and the picturesque. Waterfalls remained popular subjects, and the 'discovery' of many of the world's great cataracts gave further encouragement to writers and artists. Reports from North America about a huge cataract on the Niagara River began to circulate in the seventeenth century. The earliest eyewitness account of Niagara Falls we have is that of Father Louis Hennepin. This Flemish missionary and explorer was a member of a French expedition that arrived there in December 1678 and established a fort nearby. Starting with his *Description de la Louisiane* (1683), Hennepin published several accounts of Niagara Falls. In his book *A New Discovery of a Vast Country in America*, first published in French in 1697, he described the falls as 'the most beautiful and altogether the most terrifying waterfall in the universe'.[9] It appears that Hennepin recognized both the beautiful and the sublime in the great waterfall, though the latter aesthetic concept did not become a major subject of philosophical debate until the eighteenth century.

The illustration accompanying Hennepin's accounts of Niagara is regarded as the first known picture of the Falls. Three versions of the print were published in various editions of the missionary explorer's books. In his descriptions of Niagara Falls, Hennepin greatly exaggerates the height of the cataract, something reflected in the accompanying illustration. Some writers have commented that this image is so wrong it could not have been based on any sketch or drawing done on the spot, but if the lower three-quarters of the falls in the print are covered, a fairly accurate representation of the broad cataract emerges. A much more accurate representation of Niagara Falls is Richard Wilson's *View of Niagara, with the Country Adjacent*, a painting based on a drawing made on the spot in 1768 by British army officer Lieutenant Pierie. Wilson's painting was the beginning of a flood of Niagara pictures, the falls viewed from a variety of positions and under various conditions, such as day and night,

The first known published image of Niagara Falls, from Father Louis Hennepin's *Nouvelle decouverte d'un très grand pays situé dans l'Amerique . . .* (1697).

summer and winter. This was further encouraged by the growth of tourism, which was facilitated by improved transportation.

Growing appreciation of the American landscape, reflecting both European Romanticism and US nationalism, contributed to the emergence of the Hudson River School of artists, who were active from about 1825 to the early 1880s. Several of these artists were originally from Europe, including Britain, and they brought with them the popular taste for waterfalls. A favourite subject was Kaaterskill Falls, a famous beauty spot within easy reach of the major urban centres of the northeastern USA. Among the artists who painted this picturesque cascade were Asher B. Durand (1796–1886), born in nearby New Jersey, and the Englishman Thomas Cole (1801–1848).

With the advance of the American Frontier during the nineteenth century, more of the continent's dramatic landscapes were revealed to European eyes, providing new subjects for adventurous artists. Some of these were themselves explorers. Englishman Thomas Moran was a member of the 1871 Hayden expedition to Yellowstone, and his painting *The Grand Canyon of the Yellowstone* (1872) gives prominence to the magnificent falls there. Among the newly 'discovered' waterfalls that inspired many nineteenth-century artists were those in the Yosemite Valley.

opposite:
Hermann Herzog, *Mountain Landscape and Waterfall*, 1879, oil on canvas. Born in Germany, Herzog eventually settled in the United States.

Thomas Cole, *Falls of the Kaaterskill*, 1826, oil on canvas. This unusual view is framed by a rock arch in the cliff behind the waterfall.

Thomas Ayers's painting *The Yo-Hamite Valley* (1855) was one of the first images of Yosemite to reach the East Coast, the now famous view including Bridal Veil Falls. Yosemite Valley is particularly associated with Albert Bierstadt, whose canvases include images of Yosemite Falls, Bridal Veil Falls and Nevada Falls. While one of the characteristics of the Hudson River School was

Thomas Cole, *Kaaterskill Falls*, 1826, oil on canvas.

Albert Bierstadt, *Lower Yellowstone Falls,* 1881, oil on paper on canvas. Biersadt became well known as one of the 'Hudson River School' painters.

its realism, another was its emphasis on the sublime, something that led some painters to use considerable artistic license in their depiction of mountain scenery. For example, Thomas Hill depicted the Yosemite Valley as much narrower than it really is, thus emphasizing its sublimity.

Some artists of the Hudson River School travelled far beyond the United States. While Frederic Church (1826–1900) became well known for his paintings of North American landscapes, notably Niagara Falls, his pictures also record his extensive travels overseas. Among his most famous paintings is *The Heart of the Andes* (1859), which depicts a mountain valley with snow-capped peaks in the far distance, the foreground enlivened by a waterfall.

A picture frequently compared with this work by Church is *Land of the Lotus Eaters* (1861), painted by the African American artist Robert S. Duncanson (1821–1872). This painting, too, features waterfalls, one dominating the middle ground of the composition. Sharon Patton describes this painting as 'an escapist landscape'.[10] Born in an age of slavery, Duncanson worked largely as a painter of abolitionists' portraits and landscapes, several of which feature waterfalls as the principal subject. Escape was, understandably, something very much on the minds of African Americans in a slave society. Another African American painter, Grafton Tyler Brown (1841–1918), made his escape by joining the pioneers out West on the eve of the Civil War. One of his landscapes, *Grand Canyon of the Yellowstone from Hayden Point* (1891), recalls Hayden's picture of the same scene, with its splendid waterfall, painted nearly twenty years earlier. As Patton explains,

> The dramatic scenes he [Brown] depicted were places where few blacks actually lived and those few did so virtually free of racism. While these landscape views reflected whites' ideas of nationhood, for African-Americans they were ironic images about freedom and a process of cultural colonization.[11]

Australia

While the Hudson River School was flourishing in America, a similar artistic response to landscape was occurring in Australia, where European exploration and settlement were proceeding rapidly. For a continent notable for its generally low relief and widespread aridity, Australia has a remarkably large number of waterfalls, most of them found in the mountain ranges and plateaus close to the east and southeast coasts. It was in the southeast of the country and along the Pacific coast that European settlement began in the late eighteenth century and where the major centres of population developed. The main exception was the Swan River region of Western Australia. For many of the European explorers and early settlers, much of the Australian landscape, with its extensive stretches of apparently featureless flat plains, scrub vegetation and forbidding desert, was unattractive, even repellent. In contrast, Australia Felix, with its temperate climate and relatively abundant rainfall, as well as its forested mountains and river valleys, was seen to possess considerable scenic beauty. Here, as in America, artists sought to capture this beauty on paper and canvas.

Waterfalls, so very appealing to European taste at the time, were understandably popular subjects. Not surprisingly, the falls within or easily accessible from the main areas of European settlement were those that were most often drawn and painted. Hence, there are many nineteenth-century pictures of waterfalls in the Blue Mountains, easily reached from Sydney, and of the Wannon Falls, midway between Melbourne and Adelaide in an area settled by pastoralists in the 1830s. In the nineteenth century, the latter waterfall was painted by artists including Louis Buvelot (1814–1888), Nicholas Chevalier (1828–1902) and Thomas Clark (c. 1814–1888).

Among the Australian waterfalls most recorded by artists is Wentworth Falls in the Blue Mountains, a high but often slight cascade. In 1825 it was visited by members of a French scientific expedition, one of whom, the artist E. B. de la Touraine, made an engraving of the brink of the falls and the view beyond. Probably the most important artist to paint Wentworth Falls in the first half of the nineteenth century was the English artist Augustus Earle (1793–1838), who worked in Australia and New Zealand between 1825 and 1828. Other nineteenth-century artists who depicted Wentworth Falls include John Skinner Prout (1805–1876), Eugene von Guèrard (1811–1901) and William Piguenit (1836–1914).

Govett's Leap, with its high, slender waterfall, was another popular Blue Mountains scene, one painted by artists like Eugene von Guèrard, William Raworth and William Leigh (not to be confused with his famous American namesake noted for his

'Wild West' paintings). Among those who painted Blue Mountains landscapes was Conrad Martens (1801–1878), who, like Augustus Earle, had sailed with Charles Darwin on the *Beagle* expedition. Born in England, his father a German migrant, Martens arrived in Sydney in 1835, making his home in that city until his death in 1878. Many of Martens's Australian paintings are rural scenes in settled areas, but his subjects also include mountain scenes, some with waterfalls. His painting *Fitzroy Falls* is an early example of the artist's Australian landscapes. His watercolour of *Apsley Falls* (1874) is a work that was commissioned

Louis Buvelot,
Upper Falls on the Wannon,
c. 1872, oil on canvas.

opposite:
Augustus Earle,
Bougainville Falls,
Prince Regent's Glen, Blue
Mountains, c. 1838, oil on
canvas. This waterfall later
became known as
Wentworth Falls.

Conrad Martens,
Fitzroy Falls, c. 1876,
watercolour.

by the newly established New South Wales Academy of Art when Martens was at the height of his fame.

Twentieth-century painting

In painting as well as in literature, waterfalls, though no longer the popular subject they had been, were not entirely neglected by twentieth-century artists. During the first two or three decades of the century, many artists who had come to prominence earlier remained active and popular with the general public. Among the most important was Thomas Moran. His paintings influenced the us Government in its decision to establish the American national park system in 1916. Waterfalls in the Yellowstone region were among the subjects he chose for his paintings. In 1900 he painted one of his most dramatic pictures, *Shoshone Falls*, an impressive Idaho cataract then known as 'the Niagara of the West'. Active well into his eighties, Moran continued to paint landscapes, including his beloved Yellowstone. While becoming less fashionable among the art cognoscenti, waterfalls remained in favour with the general public, whose sentimental tastes were catered for by the likes of Maxfield Parrish (1870–1966). This American painter's gorgeously coloured landscapes, often adorned with lightly though modestly dressed pretty ladies, were reproduced in their hundreds of thousands on magazine covers, calendars and prints.

Welshman James Dickson Innes was one of the early twentieth-century artists who retained an interest in themes long popular with landscape painters and their public, but whose new stylistic freedom brought a freshness to the subject. Two waterfall paintings of his can be seen in London's Tate. While the more progressive artists of the late nineteenth and early twentieth centuries had generally rejected picturesque and sublime landscapes, painters including the 'Primitive' Henri Rousseau (1844–1910), Expressionists Wassily Kandinsky (1866–1944) and Franz Marc (1880–1916) and Fauve Henri Matisse (1869–1954), all did waterfall paintings, albeit with varying degrees of abstraction. 'Precisionist' Georgia O'Keefe (1887–1986) painted pictures of waterfalls in

Franz Marc,
The Waterfall, 1912,
oil on canvas.

New Mexico and the Hawaiian Islands. Mention should also be made of the American stained-glass artist Louis Comfort Tiffany (1848–1933), whose Art Nouveau windows depicting landscapes typically feature waterfalls. Arshile Gorky (1904–1948), an American abstract impressionist painter born in Armenia, produced a series of pictures based on natural forms, including his *Waterfall* (1943), now in the Tate Collection.

Waterfall music

From the *Idylls* of Theocritus to the onomatopoeia of Southey's
'Lodore', human creativity responded to the natural music created
by waterfalls, their various and varying sounds part of their appeal
to the senses. These natural sounds inspired both poets and
musicians, probably since the dawn of humanity. Even among
non-literate tribal societies today, sounds of nature, including
waterfalls, inspire song. Steven Feld has studied the musical tradition of the Kaluli people in Papua New Guinea who compose
their songs beside streams and waterfalls, 'singing with and to
them'. For the Kaluli, 'composing songs is like getting a waterfall
into your head'.[12]

In the music of China, as in its painting and poetry, the
landscape has long been a source of inspiration, and for over two
thousand years Chinese musicians have evoked waterfalls in
their playing. One of the most famous of these compositions is
Liu Shui (*Flowing Water*), originally paired with another piece,
Gao Shan (*High Mountain*). This landscape-inspired music is
said to have been composed by Bo Ya, a celebrated master of the
qin or guqin, an unfretted zither. *Liu Shui* evokes a mountain
stream descending from its source in the mountains, starting as
a mere rill, then culminating in a powerful waterfall. This piece
was already ancient when written down in tablature notation in
the *Shengqui Mipu* (*Spiritual and Mysteries Score*), a collection of
pieces produced in 1425.

Despite their popularity in European painting and literature,
waterfalls, unlike mountains and forests, have been largely ignored
in Western music, even that of the Romantic period. While the
waterfall-rich landscapes of Scandinavia were doubtless sources
of inspiration for Grieg and Sibelius, direct musical connections
with waterfalls are rare. Schubert's Symphony Number 9 in c
major ('The Great') is associated with Bad Gastein, an Austrian
town divided by a waterfall, but although the scenery of the area
probably inspired the young composer, there appears to be no
specific link between his musical composition and the cascading
river. Many minor nineteenth-century works have the word

Kuo Hsi, *Early Spring*
(detail), 1072, ink and
colour on silk.

'Waterfall' in the title, but few, if any, are well known. One that has been recorded commercially is *Echoes of a Waterfall* by the Welsh composer John Thomas (1828–1913). Written for the harp, this is an exquisite short descriptive piece, the delicately cascading notes of the rapidly plucked strings suggestive of a graceful fall.

Ancient music was given a space age role when a recording of *Liu Shui* began to be broadcast from the Voyager satellite launched by the USA in 1997. Thus music inspired by waterfalls on Earth over two thousand years ago is now voyaging through the universe.

8 Waterfalls and the Creative Mind: New Directions

'The timeless appeal of waterfalls is reflected in song and story.'
Richard Pearl, *Waterfalls: An Appreciation*[1]

Waterfalls through the lens

In the nineteenth century, the world of art and the role of the painter were challenged by rapid advances in photography, especially with the discovery of the daguerreotype process in 1839. The growing demand for landscape images to illustrate books of travel and exploration was increasingly being met by photographers, and painting was now freed to develop in new directions. The 1840s saw a great expansion of photography both as an art form and as a commercial business. Just as the Grand Tour had stimulated demand for landscape paintings and prints, the growth of nineteenth-century tourism created a market for daguerreotype views, particularly when tourists themselves were included in the picture. In 1845 William and Frederick Langenheim, brothers who ran a portrait photography business in Philadelphia, made a number of large panoramic pictures of Niagara Falls, each made up of five separate views. In 1853 Platt D. Babbitt obtained the monopoly for photography on the United States side of Niagara Falls. Babbitt's success in capturing the falls and including human figures was largely due to the bright reflection of light from the water, which allowed him to use an almost instantaneous exposure. These photographs were in demand as tourist souvenirs. Although daguerreotypes could not be reproduced directly, like paintings and drawings, copies of them could be made as engravings or lithographs. In this way, daguerreotype images were reproduced and used to illustrate books.

A major watershed in the development of photography was the albumen print, which made possible the reproduction of

William Henry
Jackson, *Yosemite Falls,
California, c.* 1898,
photograph.

images. This led to another innovation, the news photo, pio-
neered by Roger Fenton (1819–1869), famous for his pictures of
the Crimean War. Most of Fenton's pictures were landscapes,
however, many of them river scenes. Some of the earliest photographs of British waterfalls were by Fenton. At this time many
photographs were taken as stereoscopic pairs, used in the newly
popular stereopticon or stereoscope, which was to be found in
many middle-class homes in the late nineteenth and early twentieth centuries. Among the best-selling subjects for this home
entertainment were scenes from the Holy Land and Niagara

Falls. The nineteenth century also saw important developments in the postal service, including the introduction of postage stamps and postcards, both of which began to feature images of landscapes, including waterfalls, in the 1890s.

These were the early years of cinematography, a process by which motion, such as that of falling water, could be recorded on film and projected on the screen as a moving image. Colour photography, too, was making advances. In 1904 the Lumière brothers patented their colour screen process, starting the commercial production of Autochrome plates at their Lyons factory in 1907. This was also the time when Ansel Adams (1902–1984), was growing up in San Francisco; he was to become famous in

Roger Fenton, *Wharfe and Pool, Below the Strid*, 1854, photograph.

CARRINGTON FALLS, Robertson.
Series 32—Moss Vale.

Kerry (Copyright) Sydney.

Carrington Falls,
New South Wales,
c. 1908, postcard
from a colour-tinted
photograph.

the USA as a landscape photographer, many of his superb black
and white images featuring waterfalls.

While the skilful photographer may be able to capture something of a waterfall's rapid, turbulent movement, the resultant
image remains static. With the development of cinematography,
it became possible to record and show the movement of the
tumbling water, the rising spray and the glittering light. The
addition of recorded sound enhanced the film-maker's ability to
convey to an audience the experience of waterfalls, which began
to feature in dramas from the early days of the silent screen.

Waterfalls in film have a long history of association with sex
and violence. Filmed in Jamaica in 1915–16, the silent movie *A
Daughter of the Gods* stars Australian swimming champion and

'diving Venus' Annette Kellerman who, in the words of a *New York Times* reviewer, 'wanders disconsolately through the film all undressed and nowhere to go'. Subtitled 'Novelties of Nudity and Natation Added to the Wonders of Spectacle in an Undramatic Photo-Fable', the generally unfavourable review concedes that the film contains some beautiful tropical scenery.[2] Some of the scenes were shot at Roaring River Falls. Another early film of the silent era in which a waterfall scene is used as an excuse for exhibiting female nudity on screen is *Back to God's Country* (1919), one of Canada's first feature movies. In this film the audience is treated to a glimpse of the star, Nell Shipman, bathing, apparently naked, at the foot of a waterfall.

Still from Herbert Brenon's *A Daughter of the Gods* (1916), starring Annette Kellerman.

Among the most famous of all silent movies is D. W. Griffith's *Way Down East* (1920), a film in which the climactic scene shows the heroine, played by Lillian Gish, in dire peril as the ice of a frozen river breaks up and is swept by the current over a raging

waterfall. With the advent of sound in the cinema, audiences could now hear as well as see the waterfalls shown on the silver screen. The early talkie *Trader Horn*, filmed on location in Africa, includes a spectacular scene shot at Kabalega, or Murchison Falls. The same footage was later reused in the movie, *Tarzan Escapes* (1936), starring Johnny Weissmuller. These falls on the upper Nile were again used as a setting for action in the 1950 Technicolor film *King Solomon's Mines*. The soundtrack of the Technicolor picture *Niagara* (1952) captured the roar of the mighty cataract in a film drama in which the femme fatale betrayal by screen sex symbol Marilyn Monroe leads not only to her own murder, but to the death of her lover and to her husband's being swept over the falls. More recently, waterfalls have played important roles in films such as *The Mission* (1986), parts of which were shot on location at Iguassu Falls, and *Possession* (2002), in which there are romantic scenes at Thomason Foss, in Yorkshire. Yorkshire waterfall settings were used in the film *Robin Hood – Prince of Thieves* (1991), with Kevin Costner in the starring role. In this screen version of the story, the outlaw hero's encounter and combat with Friar Tuck were filmed at Aysgarth Falls, while in another scene Robin bathes in a pool beneath Hardraw Force. Both of these waterfalls are in Wensleydale, a considerable distance from Nottingham, Sherwood Forest and other locations regarded as Robin Hood's traditional haunts, places where one would search in vain to find scenery of this kind.

In many other films waterfalls are used as romantic settings or hair-raising hazards in stories of adventure and romance, many of them, like *Possession*, based on novels or short stories. Among these are *King Solomon's Mines* (1950), based on the book by H. Rider Haggard (1885), *Greystoke: The Legend of Tarzan, Lord of the Apes* (1984), inspired by Edgar Rice Burroughs's stories, and *The Last of the Mohicans* (1992), based on James Fenimore Cooper's book of that name. The 'jungle' waterfall scenes in *Greystoke* were filmed on location in Cameroon, while Hickory Nut Falls, featured in the climactic fight scene in *The Last of the Mohicans*, can be visited at Chimney Rock Park, North Carolina. *Indiana Jones and the Kingdom of the Crystal Skull* (2008), which includes scenes

A poster for *Niagara* (1953), starring Marilyn Monroe.

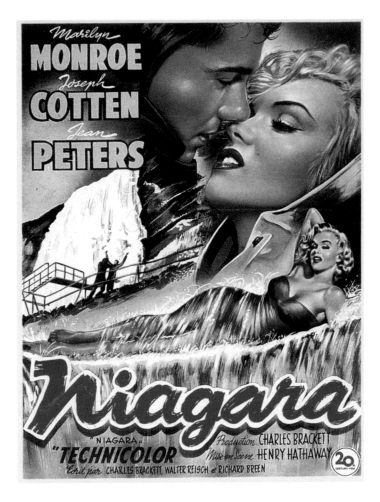

shot at Iguassu Falls, illustrates the continuing role of waterfalls in adventure films. Interest in waterfalls among film-makers is international. The award winning *Akame 48 Waterfalls* was a box office hit in Japan where it was made in 2003. Bollywood, too, has recognized that waterfalls and sex can make a powerful screen combination. Raj Kapoor's film *Ram Teri Ganga Maili* (1985) has its heroine, Mandakini, bathing under a waterfall in a see-through white sari that leaves little to the imagination.

Waterfalls are also much used in film dramas as hazards that put heroes and heroines into excitingly dangerous situations. A

Buster Keaton in the comedy adventure *Our Hospitality* (1923).

plunge over or into a waterfall, or a narrow escape from this fate, is so commonplace in adventures on the screen that it has become hackneyed. Human actors are not the only ones to thrill audiences in this way. Since 1943 Lassie has been the canine star of both cinema and TV screens, the ageless dog going from one adventure to another over the years. The 1994 film *Lassie* has the remarkable dog perform a dramatic rescue at a waterfall, emerging from a death-defying plunge over the falls looking freshly groomed.

In animated cartoon films, too, waterfalls appear frequently. Sometimes they are just part of romantic or picturesque scenic backgrounds, as in the opening scene of the 1942 Walt Disney film *Bambi*. Disney's first full-length animated feature film, *Snow White and the Seven Dwarfs* (1937), also has a scene with a waterfall, visible as the little men 'Heigh-Ho' their way home from work in the mines. In a later Disney cartoon film, *Robin Hood* (1973), a waterfall is featured in the romantic love scene between the hero and Maid Marion, both portrayed as humanoid foxes. There are instances where falls and rapids play an important part in the story, often as hazards that endanger the hero or heroine. In Disney's 1931 *Silly Symphony* cartoon 'The Ugly Duckling', the protagonist rescues some chicks from disaster as their coop is swept towards the brink of a waterfall. Similarly, in the 1933 Popeye film *Season's Greetinks*, the strongman's girlfriend Olive Oyl comes very close to tumbling over a high waterfall. To modern eyes, the representation of falling water in old cartoon films such as these is unconvincing. Even the *New York Times* reviewer of Disney's *Bambi* thought this worthy of comment: 'A waterfall that does not ripple with complete realism tears apart the illusion of a naturalistically contrived forest.'[3] In recent years the production of animated movies has involved the use of computer simulation techniques and the depiction of waterfalls on screen is now much more realistic.

A more recent Disney animated film, *The Emperor's New Groove* (2000), uses the waterfall hazard cliché to humorous effect when two characters find themselves bound to a rotting tree trunk that falls into a river. As the helpless pair are swept along by the current, they exchange the following words:

> Pacha (*facing downstream*): Uh-oh.
> Kuzco (*facing upstream*): Don't tell me: we're about to go over a huge waterfall.
> Pacha: Yep.
> Kuzco: Sharp rocks at the bottom?
> Pacha: Most likely.
> Kuzco: Bring it on.

Waterfalls on the printed page: adventure, romance and fantasy

In the world of fiction, perhaps the most famous incident at a waterfall is that in which Conan Doyle's redoubtable detective, Sherlock Holmes, struggles with the criminal mastermind Professor James Moriarty on a narrow rocky ledge above the Reichenbach Falls. When Conan Doyle wrote this scene, the Swiss waterfall had long been a celebrated tourist attraction. Here is how Holmes's trusty companion, Dr Watson, describes the dramatic location in 'The Adventure of the Final Problem', published in the *Strand Magazine* in 1893:

It is, indeed, a fearful place. The torrent, swollen by the melting snow, plunges into a tremendous abyss, from which the spray rolls up like the smoke from a burning house. The shaft into which the river hurls itself is an immense chasm, lined by glistening coal-black rock, and narrowing into a creaming, boiling pit of incalculable depth, which brims over and shoots the stream onward over its jagged lip. The long sweep of green water roaring forever down, and the thick flickering curtain of spray hissing forever upwards, turn a man giddy with their constant whirl and clamour. We stood near the edge peering down at the gleam of the breaking water far below us against the black rocks, and listening to the half-human shout which came booming up with the spray out of the abyss.[4]

As Sherlock Holmes aficionados know, despite the apparent demise of the hero along with Moriarty in this struggle to the death, the wily detective lived to have many more adventures, which were published in the *Strand Magazine* until 1927. The *Strand* was one of many popular monthly magazines that flourished round the turn of the century, their very readable content enlivened by illustrations. Sidney Paget's full-page drawing of 'The Death of Sherlock Homes', which illustrated 'The Final Problem', shows the detective and Moriarty in mortal embrace, teetering at the cliff edge above the chasm into which the Reichenbach crashes.

In the popular press, the drawn image became even more important with the emergence of comic magazines and their offshoot, the newspaper comic strip, in the late nineteenth and early twentieth centuries. First appearing in 1936, Lee Falk's masked Phantom remains one of the most successful comic strip and comic book superheroes. For the Phantom, a waterfall marks a refuge from the dangerous world in which he strives against evil. To reach the Phantom's hideaway in the Deep Woods, it is necessary to pass through the waterfall that hides the secret entrance, a passage through the rocks which opens onto a glade in front of the Skull Cave, home of generations of Phantoms,

Sidney Paget, 'Sherlock Holmes struggles with Professor Moriarty at the Reichenbach Falls', illustration published in the *Strand Magazine* (1893).

'The Ghost Who Walks'. These landscape details are faithfully included in the 1996 film *The Phantom*, which starred Billy Zane as the masked hero.

In the realm of popular literature, fantasy underwent a late twentieth-century resurgence. This owed much to the literary and commercial success of the fictional works of J.R.R. Tolkien (1892–1973), today a multimillion-dollar publication and film production industry. There developed a flourishing pictorial art business that provides colourful dramatic images used as book illustrations, on book covers and on other merchandise marketed to Tolkien devotees. A large proportion of these images are of landscapes, the picturesque, often sublime fictional settings of Tolkien's stories. The landscape of Tolkien's imagination is very diverse; parts are mountainous with tumbling streams that form waterfalls. In the first chapter of *The Hobbit* (1937), the first of the Middle Earth novels, the central character, Bilbo Baggins, excited by the songs of his unexpected Dwarvish visitors, feels an urge for adventure and wishes 'to go and see the great mountains, and hear the pine-trees and the waterfalls'.[5] One of Tolkien's verses, in a song sung by Tom Bombadil in *The Lord of the Rings*, contains the line, 'O wind on the waterfall, and the leaves' laughter'.[6] Clearly, waterfalls were seen by Tolkien and his characters as appropriate features in a landscape of romance and adventure. In the tales of Tolkien, waterfalls are encountered on several occasions, some of them being named, notably the Falls of Rauros in *The Lord of the Rings*, and the Falls of Sirion in *The Silmarillion*. In *The Hobbit* and *The Lord of the Rings*, Rivendell is an important setting, a hidden valley approached through mountains incised by deep ravines and gullies containing waterfalls. Tolkien's painting of Rivendell, used as an illustration in the second printing of the 1937 edition of *The Hobbit*, includes a waterfall. Later illustrators, such as Alan Lee and David Wyatt, have made more dramatic use of waterfalls in their paintings of this spot, and have created impressive images of other falls described in Tolkien's works. Waterfall paintings by these and other artists are found inside and on the covers of books by and about Tolkien.

Waterfalls have continued to play an important role in the flood of fantasy literature that followed the success of Tolkien. Artist and writer James Gurney created Dinotopia, a fictional island co-inhabited by dinosaurs and people of a lost civilization. His *Dinotopia: A Land Apart from Time* (1992) and *Dinotopia: The World Beneath* (1995), which capitalized on the popular interest in prehistoric monsters and fantasy worlds, also reflected the common appeal of waterfalls. Dinotopia's Waterfall City is an imposing place sited, improbably, amid enormous cataracts that, from Gurney's 1988 painting, appear to be a combination of Niagara, Victoria and Iguassu Falls.[7]

The focus on popular literature and art in the previous chapter dealing with the period from late Victorian times to the turn of the twenty-first century does not imply that, in the world of 'high culture', waterfalls were ignored. While the decline of Romanticism and the rise of Modernism probably did tend to remove waterfalls from favour among 'serious' writers and painters, these landscape features did not vanish completely from their works. Thomas Hardy (1840–1928), sometimes considered as the last Victorian novelist and the first Modernist poet, occasionally used waterfall settings. In his poem 'Under the Waterfall', Hardy evokes two lovers sharing a picnic beside a waterfall. There are several references to waterfalls in Hardy's first published book, the melodramatic novel *Desperate Remedies* (1871), and falls are mentioned elsewhere in his writings. In *A Group of Noble Dames* (1891) a coastal waterfall is described in some detail, the author noting that the place was a popular beauty spot, the subject of many paintings and photographs. Here, as we often find in his work, Hardy draws our attention to the changing Victorian world, in this case the growth of tourism in rural and coastal areas, and technological advances in photography. The two developments were, in many ways, interconnected. The touristic exploitation of waterfalls is more directly addressed in a poem by the Australian writer Sydney Church Harrex (b. 1935), 'Walking the Waterfall at Ocho Rios'.[8] This is about a visit to Dunns River Falls, a scenic but highly commercialized Jamaican tourist attraction where, under the supervision of local guides, visitors climb the waterfall that cascades down to the beach below.

Harrex dedicated this piece to Wilson Harris (b. 1921), the highly acclaimed novelist whose books clearly reveal the influence of the landscape of his native Guyana. This South American country abounds in spectacular waterfalls, a fact reflected strongly in Harris's first novel, *Palace of the Peacock* (1960). Insofar as this opaque work has a plot, it is the tale of a group of men who, in search of a lost Native American community, travel by boat up

a river, experiencing hazards, violence and death along the way. When the surviving crew reaches the foot of a huge waterfall, the three men abandon their boat to climb the great cliff over which the river tumbles, falling to their deaths in the attempt.[9]

Another major twentieth-century writer to have made dramatic use of a waterfall as a barrier and place of violence is William Golding. The theme of his novel *The Inheritors* (1955) is that modern humans outlasted Neanderthals through superiority in the use of violence. Migrating from the coast to their upland summer hunting-ground, a group of hominids make camp at a naturally protected site on a cliff terrace beside a tremendous waterfall. Here they are attacked by some 'new people'. The latter perform a religious ceremony before attempting to bypass the waterfall by making their way with their boats up the cliff and through the terrace site occupied by hominids. The waterfall features prominently in the violent encounters between the two groups and the human sacrifices performed by the newcomers.[10]

Waterfalls play a different but equally dramatic role in a late novel by Austrian writer Heimito von Dorder (1896–1966), *Die*

Tourists climbing Dunns River Falls, Ocho Rios, Jamaica.

Wasserfälle von Slunj or *The Waterfalls of Slunj* (1963). The story is set mainly in the Vienna of the Austro-Hungarian Empire, but two important sequences occur at the waterfalls on the edge of the Croatian town of Slunj. First, an English couple visit this beauty spot while on honeymoon, then later in the novel, frustrated in love, their son dies at the same place. Here we see again the common association between waterfalls and the human experience of love and death.[11]

The word 'waterfall' appears quite frequently in the titles of novels, *The Waterfall* being the name of several works of fiction. The most highly acclaimed of these is Margaret Drabble's (1969). While the book does include a waterfall scene and elsewhere a character performs a 'waterfall' card trick, the novel's title is metaphorical, having the sexual connotatio discussed in chapter Five. As we have seen, some twentieth-century poets wrote about waterfalls; the Americans Samuel Menashe (1925–2011) and David Wagoner (b. 1926) both composed pieces titled simply 'Waterfall'. Wagoner's poem appears to have been inspired as much by his wonder at the hydrological cycle as by the sight of a tumbling stream, reflecting, perhaps a growing awareness

Tourists at Niagara Falls, 19th century.

of the processes of nature that maintain life on the planet: 'the same water, having been meanwhile / Everywhere under the moon, salted and frozen, / Thawed and upraised . . .'. With its economy of words contrasting markedly with the waterfall poetry of the Romantic age, Menashe's typically short poem will serve as an example of a twentieth-century descriptive piece:

Water falls
Apart in air
Hangs like hair
Light installs
Itself in strands
Of water falling
The cliff stands

Waterfall music

Perhaps reflecting a recent revival of artistic interest in waterfalls, two contemporary Australian composers have written pieces associated with them. Rosalind Carlson's *Waterfall in Spring* (1997) is written for flute and piano ensemble, while Nigel Westlake's 'Our Mum was a Waterfall', recorded in 1987, takes its odd title from a poem. Artificial falls, in the form of fountains, have received attention from composers, including Resphigi (*The Fountains of Rome*), and Liszt (*Fountains of the Villa d'Este*). Ravel's *Jeux d'eau* is often translated as 'Fountains'.

Waterfalls may seem unlikely material for jazz, especially hard bop, the kind of music that was performed by Art Blakey and the Jazz Messengers, but in 1982, in the last session in which trumpeter Wynton Marsalis recorded with this group, his composition 'Waterfalls' was one of the numbers played. Crossover jazz–New Age saxophonist John Klemmer also composed and recorded a piece with the title 'Waterfall'. It is in the world of late twentieth-century popular music that waterfalls appear to have gained greatest favour, if only as metaphors. Best known are Paul McCartney's 'Waterfalls' (1980) and the identically named song recorded by the group TLC for their very successful album *Crazy Sexy Cool* (1994). Other waterfall song titles include 'Silver Waterfalls' and 'Cascade', both recorded by Siouxsie and The Banshees and released in 1991 and 1989 respectively, and 'Screaming Waterfalls', recorded by 8 Storey Window and released in 1995. The disc sleeve for Paul McCartney's 'Waterfalls' is unusual for a popular music CD in that it has a picture of a waterfall, in this case a surrealistic image in which the falls are cupped in a pair of hands above a range of snow-covered mountains. More naturalistic waterfalls, usually simple photographic images, are commonly found on the covers of discs of New Age and relaxation music, the latter often recorded over a background of sounds of nature – waves breaking on the shore or a rushing stream, for example.

Waterfall paintings

Some post-war painters found inspiration in waterfalls. Influenced by action painting and the minimalist and conceptual movements as well as Chinese landscape painting and Romanticism, American artist Pat Steir (b. 1940) produced a series of black and white waterfall paintings in 1990–91. Steir applied the paint in a way that allowed it to stream down the canvas, representing both the medium and the waterfall itself. Very different are the more or less naturalistic images created by contemporary Australian artist Jeffrey Makin, who has painted over 30 of the waterfalls of his native country. Some Australian Aboriginal artists, too, have found inspiration in these features of the natural landscape. Joe Alimindjin Rootsey (1918–1963) painted several pictures of waterfalls in the 1950s. A more recent work, *Waagle – Rainbow Serpent* (1983), a figurative painting by contemporary Aboriginal artist Shane Pickett, has a cascading waterfall in the background. It is a twentieth-century reminder of the ancient Dreamtime myths that associate the Rainbow Serpent with waterfalls.

With the twentieth century now well behind us, there are signs that waterfalls retain interest among contemporary painters. In 2006 an oil on canvas landscape, *Before Vermeer's Clouds* by

In the advertising world, waterfalls symbolize nature and purity.

Martin Greenland, won Liverpool's John Moores Prize with its distant view of a town featuring a waterfall in the foreground. The picture represents the artist's vision of heaven.

No account of waterfalls in twentieth-century art can be complete without mention of Escher's famous *Waterfall* lithograph (1961). Born in the Netherlands, Maurits Cornelis Escher (1898–1972) was little known until the 1950s when his impossible imaginary scenes gained popularity. The architectural *Waterfall* is perhaps the best known of these, and is now such a popular image that it even appears on T-shirts.

Advertising the pure and natural

Waterfalls are widely regarded as symbols of purity, suggestive of pristine nature, the world as it was before its defilement by the artificial ways of modern life. These associations are commonly exploited by advertisers promoting products and services that are meant to appeal to a public which is preoccupied with cleanliness and 'loves Nature'. Consumers are supposed to be concerned about dirt, pollution, contamination and perhaps spiritual or mental health. The sad truth is that many waterfalls are by no means pure or pristine. The quality and quantity of water that flows over the falls and the nature of the immediate surroundings are likely to have been altered by human activity such as agriculture, forestry, power generation, tourism and urban development. While much of human impact on waterfalls and their environs has been detrimental to their appearance, for many

Soft drink advertisement in Brisbane, Australia.

centuries cascades and cataracts, both natural and artificial, have been incorporated into landscapes that have been designed to delight the eye. The following chapter, therefore, considers how the creative mind has made use of waterfalls in the landscape.

9 The Designed Landscape

'The waterfall's descent is carefully constructed with rocks that
create fascinating sights. To make the view more appealing and
natural, the gracefully manipulated cascade is often half screened
by vegetation.'
Francis Ya-Sing Tsu, *Landscape Design in Chinese Gardens*[1]

Sources of pleasure; fountains of delight

One important medium in the development of the creative
arts has been the land itself. From the earliest times, humans
have shaped the earth on which they lived, first, no doubt, un-
consciously as they moved about and exploited the natural
environment to satisfy their basic needs. Later, they began to
make apparently conscious efforts to modify their environment
in their struggle for survival, by the management of vegetation
with the controlled use of fire and the construction of shelters,
for example. We must be cautious, however, when we attribute
consciousness to these early human activities, and should remember the comparable behaviour of other animals, like web-making
by spiders, nest-building by birds and dam construction by beavers.
By obstructing streams in this way, beavers create low falls of
considerable width.

As discussed earlier, there seems little doubt that the aesthetic
sense in humans and their landscape preferences developed
during the millions of years of hominid evolution. There is much
empirical evidence that park-like landscapes, similar to the
tropical savanna lands in which our ancestors evolved, are widely
favoured by people today. Open grassland with scattered trees still
appeals to us so strongly that we commonly seek to recreate this
landscape in our parks and gardens. For similar reasons, we also
prefer landscapes with water.[2]

Probably the earliest human interference with nature's water
supply was for strictly practical purposes, like digging basins for
collecting drinking water and constructing fish traps. Natural

springs would have been improved by clearing vegetation, earth and stones that impeded access. In some cases, a basin would be created at the site to facilitate use. While it is usually easy to drink or to fill a container from a basin such as this, the water is certain to contain material that has fallen in, including dust and plant and animal matter of all kinds. The sediment that normally accumulates in the basin may cloud the water when disturbed, and the pool itself is likely to be the home of a variety of aquatic life. It is usually preferable, therefore, to drink or to fill a container with water at a spring that spouts forth at some convenient height. It is for this practical reason that at many spring sites all over the world the flow has been artificially channelled and made to gush out over a projecting lip or through a spout. A simple gutter or pipe made of wood, bamboo or other material is a common means of achieving this, something still to be seen in places without modern water supply systems. It seems probable that arrangements like these were commonly made by our hunter-gatherer ancestors. For the reasons already discussed, falling water thus artificially contrived would be a source of pleasure, a fountain to delight the eye and ear. Often considered sacred, and commonly believed to possess healing properties, many springs were embellished with carved stonework, some being enclosed in elaborate buildings.

Hydraulic landscapes

With the development of agriculture, and especially farming dependent on artificial irrigation, the sight and sound of cascading water would have been familiar to people who lived in areas where weirs, channels and sluices were commonplace. Among the regions where landscapes of this kind developed in ancient times were Mesopotamia in the basin of the Tigris and Euphrates Rivers, the Nile Valley of Egypt, the Indus Valley of the Indian subcontinent, north China, pre-Columbian Mexico and Peru, and the Hawaiian Islands. Similar landscapes – and soundscapes – can be experienced today, as in the terraced rice fields of much of south-east Asia. Irrigation channels that cascade from terrace to terrace are part of the scenic charm of these areas. In Europe the Romans made practical use of artificial cascades in their aqueducts. As well as conveying water quickly from one level to another and dissipating potentially destructive energy, artificial cascades also helped aerate it. This assisted the purification process.[3]

Over time, the sights and sounds of falling water would become associated with a productive environment favourable to human survival and well-being and, as habitat theory suggests, would thus give aesthetic pleasure. From this it would be a logical step to create these sights and sounds for pleasure alone. With the development of societies in which some people had the leisure and means to indulge in such luxuries, designed landscapes were created for the enjoyment of the privileged classes. In many parts of the world artificial waterfalls and fountains became part of the suite of features typically found in the pleasure grounds of the rich and powerful.

The ornamental use of cascades, fountains and waterfalls

Perhaps the earliest pleasure gardens were those of ancient Mesopotamia and Egypt, where irrigation-based agriculture developed at least seven thousand years ago. The Egyptian, Sumerian and Assyrian civilizations were supported by high levels of agricultural production made possible by large-scale irrigation systems. The constructed terraces of the Hanging Gardens of Babylon, believed to have been built by Nebuchadnezzar about 605 BC, probably used water raised from the Euphrates by mechanical means and allowed to return to the river forming ornamental cascades in its descent.[4] No trace of such a structure has yet been discovered at the site of Babylon and it is possible that the Hanging Gardens of ancient legend were actually those which have been found at Nineveh.

Later, the Greeks and Romans made fountains an important element in their towns, pleasure grounds and sacred places, combining utilitarian, religious and aesthetic functions in their development of natural springs. Typically, the water was conducted into

Roman nymphaeum at Gerasa, now Jersash, Jordan, 2nd century AD.

basins, which were often adorned with sculpture. In many places they were covered and enclosed within a building of some kind.

Imperial Rome had over 120 public fountains, as well as many others in private ownership. These were served by an extensive system of aqueducts that brought water from distant sources. Ornamental fountains, many in the form of artificial cascades, were typically found in the homes and gardens of wealthy Romans. Derived from the Greek fountain, and originally sacred to nymphs and river gods, the Roman nymphaeum was a more elaborate architectural feature that was fairly common even in private villas. It was essentially a vaulted room with fountains, cascades and sculpture which made a cool retreat on hot summer days. In suitable climates, alfresco dining was popular. The garden triclinium or outdoor dining area served this purpose well, and water commonly played an important role in the design and function of these spaces. Long verandahs, marble pavilions and vine-covered pergolas gave shade to the diners as they reclined on couches beside the water channels and pools where cascades and fountains made the summer air cool and pleasant. At Emperor Hadrian's Tivoli villa, an enormous palace complex, water was employed on a particularly large scale. There water features ranged from little fountains to gigantic waterfalls, all of which created a variety of sights and sounds.

For the desert people who developed the culture of Islam from the seventh century AD onwards, water was rarely so abundant as to allow its extravagant use. It was nevertheless a very important feature of the Islamic garden, splendid examples of which were created in parts of western and southern Asia, North Africa and southern Europe. Typically, the Islamic garden is rectilinear in layout, divided into four rectangular sections by four channels, symbolic of the rivers of blessedness, flowing from a central fountain. It is a pattern that closely resembles those often found on Persian carpets, and 'Persian garden' is a generic term commonly used for this type of landscape design. Variations on this theme can be found in Islamic gardens from the Alhambra in Spain to Isfahan in Iran and the Taj Mahal in India. The pools, fountains and cascades that are common features of these delightful open spaces testify to the skill of the designers in making the best use of an often limited supply of water.

Displays of falling water were particularly impressive in gardens created in those parts of the Islamic world where, like the Roman Tivoli, sloping terrain and abundant streams provided ideal conditions for this kind of landscape treatment. Nowhere is this better seen than in the Vale of Kashmir. Here, mainly on and around the shores of Lake Dal, hundreds of gardens were created during the reigns of India's Mughal emperors Jahangir (*reg.* 1605–27) and Shah Jahan (*reg.* 1628–58). The

A Mughal water garden in Kashmir with *chadar* (water chute).

palace gardens were laid out on terraces over which water cascaded in broad falls or tumbled down narrow, steeply inclined
and often richly carved chutes known as *chadars*. The angle of
incline of the *chadar* is designed to reflect sunlight effectively,
while the scalloped or herringbone patterned surface of the chute
down which the water tumbles gives an attractive texture to the
cascade. Most of the Mughal water gardens have been destroyed,
but some impressive examples survive, notably Shalamar Bagh
and Nishat Bagh. At Shalamar, Jahangir gave audiences while
seated on a black marble throne set over a waterfall, an arrangement that was no doubt intended to enhance the ruler's powerful
image. A second throne is set in front of the cascade, which thus
forms an impressive background. In the garden at Achabal, a
huge cascade, over twenty metres wide, is fed by a powerful
natural spring.[5]

The eastward spread of Islam continued, extending as far as
what is now Indonesia and parts of the Philippines, but in large
areas of east Asia cultures based on older belief systems survived.
Civilizations that embraced Buddhism, Hinduism, Confucian-
ism and Taoism, like those which arose further west, showed a
similar appreciation of water in their landscape design. Typically, placid water surfaces reflect the surrounding buildings,
landscape and sky above, contributing to an overwhelmingly

calm atmosphere. Within this south-east Asian cultural region, however, ornamental cascades and fountains were created at some suitable sites.

The naturalistic waterfalls of traditional Chinese and Japanese gardens are very different from the manifestly artificial cascades and other water features so far discussed in this chapter. Whereas many of the world's designed open spaces are the work of professionals in garden design and construction, Chinese and Japanese gardens were typically created by monks, poets and painters.[6] Artificial landscapes were designed to give the impression that they were the work of nature, or at least to display many of the qualities of a natural landscape, such as irregularity and asymmetry. Hills were built, lake basins excavated, rocks introduced and planting undertaken, sometimes over an enormous area, more often on a modest domestic scale.

In all this, water played an important role, ponds, lakes and streams often occupying more than half of the total landscaped area. This may not be surprising when one remembers that, according to feng shui beliefs, the presence of water is particularly propitious. It is an idea that clearly echoes habitat theory. While in general Chinese and Japanese garden designers sought to achieve a tranquil effect, the exciting qualities of cascading water were commonly introduced, typically hidden in the recesses of artificial hills where they would not disturb the overall serenity of the landscape. As their remarkable landscape paintings and writings testify, the Chinese and Japanese have long appreciated the beauty of natural waterfalls, and these features were often artificially reproduced in their gardens.

In Japan, as in China, waterfalls are much-admired features in the landscape and were often introduced into gardens. The *Sakuteiki*, an eleventh-century Japanese book on gardening, gives detailed advice on the location, design and construction of waterfalls. Ten types are distinguished, each with a specific name: Twin Fall, Off-sided Fall, Leaping Fall, Side-facing Fall, Cloth Fall, Thread Fall, Stepped Fall, Right and Left Falls, and Sideways Fall. Some very old artificial falls continue to flow, including the famous cascade of the North Hill Villa, Kyoto, constructed in about 1224. The ideal position for a waterfall was at the edge of the garden, where it gave the illusion of a constant source of water behind. Where lack of water or insufficient available space made waterfall construction difficult, Japanese gardeners sometimes sought to achieve the affect by the artful arrangement of stones and sand. A pair of stones standing upright was commonly used to symbolize this feature. Despite several distinctive features such as this, traditional Japanese and Chinese garden design has much in common with the picturesque style of landscape architecture that emerged in late eighteenth- and early nineteenth-century Europe and which it influenced to some degree. When we remember the European concept of the picturesque and the more recent prospect refuge theory, it is particularly interesting to read the words of Francis Ya-Sing Tsu, author of *Landscape Design in Chinese Gardens*: 'To make the view more appealing and natural, the gracefully manipulated cascade is often half screened by vegetation.'[7] The same design principle is found in Japan, too. Matsunosuke Tatsui observes, 'To make the view more appealing, the naked view of the cascade is often avoided by a clump of trees planted in front which serves as a partial screen.'[8] The proportions of waterfalls were also regarded as aesthetically important; the *Sakuteiki* warns against making low waterfalls too wide, which would give the appearance of an artificial dam on the stream.[9] This concern with naturalism is also reflected in the general absence of fountains that jet water upwards, something contrary to the nature of water. There were of course exceptions, including the *shui shih* or 'hydraulic elegancies', such as the fountains on which danced balls and hydraulically operated robots created for the last emperor of the Yuan dynasty.[10]

As in other cultures, it is not only the sight of falling water that appeals to the Chinese and Japanese, but the sound, too. 'The different sound that single and multiple falls make is considered an important element in the general character of the area where a waterfall is positioned.'[11] While this is usually a normal part of the 'natural' experience that garden builders sought to create, there are some waterfalls in which sound is exploited in an undisguisedly artificial way. At the Carefree-Abiding Garden in

Wuxi, China, is the Rill of the Eight Musical Tones which was designed to produce musical sounds generated by water falling from different heights onto stones of different sizes, densities and shapes. Sadly, this water music ceased when the cascade was allowed to fall into neglect. Another artificial sound generated by means of falling water is that produced by the deer scarer, a simple lever device originally intended to frighten animals away from growing crops. Introduced into Japanese gardens, it is basically a pivoted bamboo pipe that, activated by falling water, repeatedly strikes a stone, generating a rhythmic, soothing sound. High in the Himalayas, the sound of bells accompanies silent Buddhist prayers that are sent heavenwards from spinning prayer wheels turned by the constant flow of tumbling mountain streams.

In the Americas, as elsewhere in the world, systems of water management played a vital role in supporting agriculture and supplying urban centres. Here, too, water was often used for aesthetic purposes, particularly in the pleasure grounds of the rich and at places of religious significance. The ancient civilizations of Meso-America reached their climax with the Maya and Aztecs, while in the Andean region of South America it was the Inca empire that displayed the greatest achievement.

In pre-Columbian Meso-America, pleasure grounds, some with water features, were typical of ruling-class houses. At some religious sites, too, artificial water features played significant roles in landscape design, including ritual pools and fountains. At sacred Texcotzingo, near Mexico City, the mid-fifteenth-century ruler of Tetzcoco built a palace with gardens on the terraced hillside. On these slopes, water flowed from a reservoir into basins from which it cascaded over rocks, making spray that, according to Aztec historian Ixtlilxochitl, fell like rain on the scented flowers.[12]

In the whole of ancient America, however, no people surpassed the Incas in the aesthetic use of water, notably in the form of cascades and fountains.[13] In addition to their magnificent agricultural irrigation systems, they built waterworks for religious and recreational purposes. Elaborate waterworks were characteristic of royal Inca estates where 'beautifully crafted fountains fed cascading waters through exquisitely carved channels'.[14]

Commonly, fountains and cascades were associated with the stone staircases that are typical features of Inca monumental sites. Outstanding examples are the splendid fountains that adorn the central staircase at Machu Picchu, and a spectacular cascade of nineteen fountains beside the staircase linking the upper part of Wiñay Wayna with its lower precinct. In a very different setting, on the shore of Lake Titicaca's Island of the Sun, there is an Inca fountain with a flight of steps and cascades on the slope above. Fountains were also notable features at Ollantaytambo, with its impressive fortress and temple overlooking the Inca town in the Urubamba valley.[15] The Inca, like people of other places and times, often oriented their buildings on features of the natural landscape, including waterfalls. At Wiñay Wayna the long flight of steps, notable for its accompanying cascade of fountains, leads to a curved building perched high above a gorge. A window at the apex of the curve frames a spectacular view of a waterfall.

An Inca four-spout fountain at Tipón, Peru.

Like the Aztecs, the Inca were defeated and subjugated by the Spanish invaders in the sixteenth century. The pre-Columbian civilizations collapsed and were replaced by a colonial system that imposed European concepts and forms on the people and landscape. Similar radical, often violent, changes were occurring in other parts of the world that were becoming increasingly dominated by Europe. It was Europe of the Renaissance that was now making its mark, bringing with it new ideas and tastes in architecture and landscape design.

A thousand years before Columbus reached America, Europe too had experienced the violent collapse of a great civilization. The disintegration of the Roman Empire had been followed by a period in which much of the achievement of the classical world was destroyed. Refinements of life, including pleasure gardens and fountains, even the more utilitarian aqueducts and irrigation systems, were destroyed or allowed to fall into neglect. The eventual rise of the Middle Ages saw a resurgence of European cultural life in which the pleasure garden and the fountain again found a place. That place was of course the world of the privileged, who lived in the castles and palaces of royalty and the nobility or in the great monasteries. Few medieval garden fountains remain today, but there are many representations of them in art, and references to them in literature and written records.[16] Medieval fountains typically took the form of an ornate column rising from a raised tank or basin sunk in the ground. The water fell from several spouts near the top of the structure that was commonly surmounted by a sculpted figure, animal or other ornamental object. Late medieval paintings often depict fountains of Gothic design, with one or more pinnacles. Stone, lead and bronze were materials commonly used in their construction.

With fountains, as with other structures and works of art, the Renaissance replaced Gothic forms with very different ones derived from the classical world of ancient Greece and Rome. It was a time when Europeans discovered a renewed interest in the arts and literature of those cultures, a period, too, when cascades and fountains became extremely popular in Europe. They were built in increasing numbers and magnificence in pleasure grounds and important public places, especially in the major centres of wealth and power. Many survive, and can be seen in the great parks and gardens of the former royalty and nobility, and in city squares. In the early years of the Renaissance period fountains often were built using classical Greek and Roman sculpture that had been unearthed from ancient sites.[17] Among the many classical features that regained popularity in garden design during the sixteenth century was the nymphaeum. Increasingly, fountains became opportunities for the lavish display of contemporary Renaissance sculpture which eventually evolved into the Baroque style. While Renaissance fountains typically featured modest flows, often descending from tiered basins on a columnar shaft engulfed by sculpture, the more ambitious creations of the Baroque period were characterized by water gushing in large quantities from a variety of orifices among masses of dramatic sculpture depicting mythological and allegorical figures. The most splendid examples are found in France, Spain, Russia, Germany and, especially, Italy. Rome has been famous for its fountains since ancient times, but the renowned 'Fountains of Rome' we know today are mainly the creation of the Renaissance, when the city's aqueducts were repaired and restored. Among the most famous of Rome's fountains are the Triton, the Four Rivers and the Bee, all the work of the eminent seventeenth-century artist, sculptor and architect Gian Lorenzo Bernini (1598–1680), and the Trevi, for which Bernini submitted a design but which was completed long after his death to the design of Nicola Salvi.

During this period, as in the days of ancient Rome, Tivoli was a favourite resort of the wealthy, who built sumptuous villas and laid out magnificent gardens there. Again water was used on a grand scale; the gardens of Tivoli, with their spectacular cascades and fountains, are among the most remarkable achievements of Renaissance and Baroque landscape design. The best known of these are the pleasure grounds of Villa d'Este. Today a popular tourist attraction, the gardens of Villa d'Este were designed for Cardinal Ippolito II d'Este by the Neopolitan architect Pirro Ligorio and completed between 1560 and 1590. To supply water to the site, a tunnel, over two metres in diameter and 600 metres

The Avenue of One Hundred Fountains at Villa d'Este, Tivoli, near Rome.

long, was excavated. This abstracted part of the Aniene river's flow and conveyed it, at a rate of 77,000 litres a minute, to a reservoir at the top of the gardens. On its descent through the villa grounds the water jets, spouts and tumbles in a profusion of magnificent fountains and cascades. The fountain-makers of the Villa d'Este used the most advanced hydraulic techniques of the day to produce varied and spectacular effects that included organs and sound-making automata activated by water-powered bellows.

Among the many other examples of Italian Renaissance and Baroque fountains and cascades are those at the Villa Lante, Bagnaia (1566), Villa Aldobrandini at Frascati (1560), Villa Garzoni at Collodi (1652) and Palazzo Reale, Caserta (1752). Italian water features strongly influenced landscape design throughout Europe, and spectacular fountains and cascades were built on a grand scale in the grounds of many palaces and mansions from Chatsworth in England to La Granja, Spain, from Vaux-le-Vicomte in France to Peterhof, near St Petersburg, in Russia. The

Baroque cascade at Wilhelmshöhe, Kassel, Germany, was over ten metres wide and about 240 metres long, yet was only a third of the length originally intended by Guernieri, who designed it in about 1700.

With the rise of Romanticism in the eighteenth and early nineteenth centuries, fountains that jetted water into the air fell out of favour, being regarded as unnatural. Because waterfalls commonly occur in nature, artificial cascades were much more acceptable. If a property contained a natural waterfall that could be incorporated into the landscape design, so much the better. At Rydal Hall, Cumbria, a nearby waterfall on Rydal Beck has been a feature of the landscaped grounds since the seventeenth century. The 'viewing house' at the waterfall site was built in 1669. Dorothy Wordsworth, who lived nearby with her brother William, described many other examples of these 'improvements' in her journals. She was often critical of the landscape designs implemented by landowners who embellished their 'pleasure-grounds' with 'pleasure-roads', 'pleasure-paths' and 'pleasure-houses' of various kinds, as well as seats, parapets, exotic trees and the like. On a tour of Scotland in 1803, Dorothy, her brother William and their friend Samuel Taylor Coleridge went to see the Falls of Clyde (see page 175). There, along the gravel footpath, they saw a variety of amenities and ornamental features that had been provided for the pleasure and convenience of visitors. These included bench seats from which to view the falls in comfort, and a 'pleasure-house' in the form of a rustic, moss-lined, circular hut, a feature 'common in the pleasure-grounds of Scotland'.[18] In her earlier journal account of a walk to some little waterfalls in Somerset, Dorothy criticized the excesses of Romantic landscape design where 'art had deformed' nature by the planting of 'unnaturalised trees' and the erection of 'ruins, hermitages, etc. etc.'[19]

In their enthusiasm for picturesque and sublime scenery, Romantic travellers were sometimes prepared to take considerable risks in order to get close to nature. Many landowners made improvements at waterfall sites but dangers commonly remained. Paths to waterfalls were often cut through rugged ravines made hazardous by unstable rocks and precipices, and some of the

Rydal Beck waterfall and Stock Ghyll Force, both near Ambleside, Cumbria, 1833.

best vantage points from which to view the falls were themselves dangerous.[20] In places like these, rustic staircases, footbridges, belvederes and even more substantial structures were erected. In some of these, mirrors were placed in such a way as to make it appear that water was falling onto the spectator.

Accessible places endowed by nature with waterfalls were thus converted into picturesque pleasure grounds for the enjoyment of both owners and visitors. Where there were no naturally occurring falls, a waterfall more or less natural in appearance might be created. One example is the cascade at Barrow Hall beside Derwentwater in the Lake District, a region that abounds in natural waterfalls. Here a stream was diverted to an excavated rock-face to create the artificial fall not far from the famous, natural Lodore Falls.[21]

Notwithstanding Romantic landscape tastes that gave preference to the more natural cascade, advances in technology, particularly improved pumping systems, encouraged fountain construction. Indeed, some fountains were built more as wonders of technological achievement than as aesthetic features of the designed landscape. The 88-metre jet of the nineteenth-century Emperor Fountain in the gardens at Chatsworth House, Derbyshire, is an example of this.

By the mid-nineteenth century, advances in drilling made possible boreholes that reached artesian water capable of supplying fountains. Under natural pressure, this water rises to the surface and spouts upwards. London's Trafalgar Square fountains were originally operated in this way, but later, when the natural water pressure declined, pumping machines were introduced. Fountains were increasingly operated by mechanical pumps, after which electricity became the usual source of power.

As the nineteenth century advanced, public parks began to proliferate in Europe and elsewhere. Fountains, waterfalls and other water features were commonly used to add beauty and interest to these popular open spaces. In many areas where the landscape had been defiled by mining, industrial waste dumping and other activities associated with the Industrial Revolution, public parks were seen as amenities that could help ameliorate the squalid environment in which much of the population had to live. Novelist Arnold Bennett (1867–1931) was familiar with one of these close to his birthplace in the Potteries, Stoke-on-Trent, observing that 'fountains and cascades babble' in the gardens created where previously there was industrial wasteland.[22] Fountains were also erected in town and city squares, many of those in Victorian Britain being designed as drinking fountains, provided in the hope that these would help reduce the consumption of alcohol among the lower classes.

At the same time, city parks were being established in North America, where landscape architecture was becoming an increasingly recognized profession. Italian and French baroque water gardens were being reproduced in the now rich and powerful usa, many of them on the estates of wealthy families, including the Biltmore estate in North Carolina, designed by Frederick Law Olmsted, and the Hearst estate in California. As late as the 1930s, an adaptation of the Italian water cascade was constructed in Meridian Park, Washington, DC.

The twentieth century saw a renewed interest in cascades as elements in architectural, landscape and urban design, not least in the USA. An outstanding example of the use of a natural waterfall in an architectural design is Frank Lloyd Wright's Fallingwater, a private residence in the hills of southwest Pennsylvania. Wright designed and built the house for the Kaufmann family in 1936–9, taking inspiration from the picturesque wooded, rocky site beside a cascading stream. Here, the Bear Run creek drops over a resistant sandstone ledge, and part of the building extends dramatically over the waterfall. Reproduced in numerous books on art and architecture, Wright's masterpiece is one of the world's most famous modern buildings. With its unique marriage of architecture and natural landscape, it is among the twentieth century's great artistic achievements.

Twentieth-century designers created an exciting variety of artificial waterfalls and fountains that became common features in urban and suburban settings, including city parks and squares, shopping malls, office and hotel forecourts and foyers, resort complexes and other developments. Seen on millions of television screens around the world, one of the most striking waterfall images in recent years was that at the opening ceremony of the 2000 Olympic Games in Sydney, Australia. In a spectacular juxtaposition of fire and water, the flaming Olympic cauldron ascended to the top of the stadium, gliding over the 77 horizontal steps of a cascade that began to flow as the ceremony reached its climax. On a more sombre note is the design of the National September 11 Memorial in New York, featuring waterfalls where once stood the Twin Towers.

Today, probably more than ever before, artificial waterfalls are found within as well as in the grounds and patios of buildings. They are common in restaurants, where the sight and sound of falling water contribute to a relaxed atmosphere conducive to pleasurable dining. The soothing influence of tumbling water has generated a considerable industry in the manufacture of water features for the home and workplace. An amazing range

of indoor waterfalls, including portable tabletop and desktop models, are now available in a variety of materials. As for the artificial waterfalls of the future, science-fiction writers may provide a clue to how these may develop. In his short story 'The Undercity', Dean Koontz describes a world of hundred-storey megalopolises where, on the 83rd level of one urban complex, there is a public hydroponics park complete with an artificial waterfall.[23]

Frank Lloyd Wright, Fallingwater, Pennsylvania, 1936–9.

10 Power and Human Settlements

'Waterfalls, more than any other landform, excited ambition and
prompted visions of potential development. These natural features
called settlements into being.'
William Irwin, *The New Niagara*[1]

Water and energy

While on honeymoon in the French Alps in the summer of 1847,
James Joule went with his new bride to the Cascade d'Arpenaz,
at Sallanches. It was science rather than romance, however, that
the Englishman had on his mind when he visited this waterfall. Armed with a large thermometer, he attempted to measure
the temperature of the water at the top and the bottom of the
fall, seeking proof of his theory that loss of potential energy
should cause water at the foot of the drop to be slightly warmer
than that at the brink high above. His results were inconclusive
because, in its long descent, the waterfall turned into a cloud of
spray, making accurate temperature measurement and compar-
ison impossible.

Joule won enduring recognition for his many scientific
achievements, including the discovery of the law of conservation
of energy, which states that energy can be neither created nor
destroyed but only changed from one form to another. This is
demonstrated by a waterfall. The water at the top of a waterfall
possesses gravitational potential energy. Going over the brink,
the stream begins to fall, accelerating under the force of gravity.
In this way, the gravitational potential energy that the water possesses at the top of the fall is converted into kinetic energy as it
descends. Ultimately, of course, all of this water power is derived
from the sun through the hydrological cycle. It is the heat of the
sun that causes the evaporation of water on the earth's surface,
drawing up the moisture from oceans, lakes, rivers, streams and
vegetation. The water vapour in the atmosphere condenses into

clouds from which precipitation, in the form of rain, snow or hail, descends back to earth, replenishing the water bodies there. Gravity pulls the water from higher to lower ground, causing springs, streams and rivers to flow towards the sea.

The energy of swiftly moving and falling water must have been familiar to the earliest hunters and gatherers who foraged beside fast-flowing and tumbling streams, fished in them and crossed them in their wanderings. Probably among the first practical uses to which this water power was put was the floating of timber downstream to a place nearer to where it was required. Primitive people soon realized that floating logs, or rafts formed by binding them together, could be used to transport goods and passengers, and harnessed the energy of the current for this purpose. For transportation purposes, however, waterfalls are obstacles, destructive of goods and people being carried downstream on the current, and barriers to those making their way upstream by haulage or other means.

Water power, waterwheels and water turbines

A primitive device that harnesses the power of falling water is the water lever. This comprises a pivoted beam with an open-ended bucket at one end and a hammer-like counterweight at the other. Operating in see-saw fashion, the machine is activated by water flowing into the bucket, the weight depressing that end of beam causing the other end, with its hammer, to rise. With the tilt of the descending bucket, the water inside spills out and consequently the weighted other end falls, delivering a hammer blow. The continuing flow of water causes this action to be repeated cyclically, an action suitable for crushing and pulverizing.

More efficient means of harnessing the energy of rivers and streams for mechanical purposes followed the invention of the wheel. There is much uncertainty about the origins of waterwheels, but one very early type is the Greek or Norse mill. In this primitive device, a horizontal 'wheel', essentially a set of blades or paddles fixed to a vertical shaft, is turned by the force of water directed onto it by a chute or race. To improve efficiency, the blades are fixed to the shaft at an angle, allowing the water to impart a greater thrust. At the top of the shaft are the millstones, also in the horizontal position. No gearing is involved, so the revolving millstone or runner turns at the same speed as that of the horizontal waterwheel to which it is joined directly by the vertical axle. This type of mill was probably developed in hilly regions of the eastern Mediterranean over two thousand years ago, later spreading throughout Europe to Scandinavia and eastward to China. While commonly used for milling grain, it could also perform other functions, including the operation of bellows for metal working. Requiring a relatively small volume of water with a modestly high fall, these mills were frequently found on mountain streams.[2]

An early type of vertical waterwheel, the noria or Persian wheel, unlike the Greek or Norse mill, activates no machinery. It is used to raise water directly from the river whose flow turned the wheel. Attached to the rim of the wheel are both the blades, against which the stream flow provides thrust when they enter the water at the foot of their descent, and pots or buckets, which fill with water when immersed at the bottom of the wheel's rotation. Raised aloft by the turning wheel, the water containers spill their contents into a trough as they tilt over at the top and begin their descent, the cycle continuing. Extensively used until recent times in areas of irrigated agriculture, waterwheels of this kind were found beside large rivers with powerful, smoothly flowing currents. They are unsuitable for tumbling streams broken by rapids and falls.

The noria is an example of an undershot waterwheel, one that is turned by the flow of water operating on the blades at the bottom of the rotating wheel. Possibly inspired by the noria was the earliest form of watermill, which had a vertical wheel with a horizontal axle. This type was associated with Vitruvius, a Roman architect and engineer of the first century BC who wrote a clear description of the mechanical device with its gear arrangement. Like the noria, the Vitruvian watermill is powered by the flow of water acting on the blades at the bottom of the wheel. Unlike the noria, the rotation of the horizontal axle causes the rotation of the vertical shaft by means of a simple gear arrangement, transmitting the power generated by the turning waterwheel to the millstones or other machinery. Much more efficient than the undershot mill is the overshot mill, where the wheel is turned by water supplied to the top. The paddles or buckets of the wheel are designed to hold water so that energy from both the momentum of the flow and the weight of the descending water is harnessed to operate the mill. Overshot mills require an abrupt fall of water with a drop at least as great as the diameter of the waterwheel. Thus mills of this kind are normally sited beside or close to streams with steep gradients. Similar to the overshot wheel is the pitchback or backshot wheel, the main difference being the direction of rotation. With the pitchback arrangement, the water falls onto the upstream side of the wheel. While this may lose the advantage of the momentum of the inflow, it gains from having the rotation assisted by the outflow. It also allows the mill to operate even when stream levels rise above axle height, a situation that would stop and possibly damage an overshot wheel.

The water required to operate the mill is diverted from the natural course at a point upstream and led along an artificial channel, known as the head-race, to the top of the waterwheel. To facilitate the diversion of water from the stream into the head-race and, often, to increase the head of water and create a storage reservoir or mill-pond, a dam or weir is built across the stream. The flow of water from the stream or mill-pond into the head-race is usually controlled by one or more sluices. Having done its work, the water flows from the bottom of the wheel as the tail-race. In some cases, the latter becomes the head-race of another waterwheel downstream, and there have been quite large mill complexes arranged serially in this way. Over the centuries, different varieties of mills and millwheels were introduced and the mechanical energy they generated put to a wide range of industrial uses.

The Industrial Revolution owed much to the wide application of water power as well as to coal-fired steam engines. During the late eighteenth and early nineteenth centuries, advances in the

Rutter Force, Cumbria. This waterfall was formerly harnessed to turn the overshot wheel of the down-stream mill. The building is now a holiday home.

manufacture of iron and steel allowed improvements to be made to the waterwheel, originally constructed of wood. This eventually led to the development of the water turbine. The very high-speed rotation that this produces is extremely suitable for the generation of electricity, and thus it was the development of the water turbine that led to the harnessing of falling water for hydroelectric power.

Hydroelectric power schemes range in scale from the small plant designed to supply power for the individual home to enormous projects involving huge dams, vast reservoirs, extensive transmission lines and the rest, constructed to meet the energy demands of great cities and regions with populations of millions. In terms of effect on the built environment and human settlements, there is a major difference between mechanical energy derived from the harnessing of falling water and hydroelectric power: the latter can be transmitted over long distances, whereas the former requires the consumption of energy at source.

Water power and human settlements

Many human settlements were originally sited close to rivers and streams for various reasons, not least of which was the need for easy access to water for drinking and a variety of domestic purposes. With the rise of agriculture, demand for fertile, well-watered land further encouraged the occupation of river valleys, in many of which irrigation was introduced. Increased agricultural production, especially with grain crops, led to the demand for more labour in the transporting and processing of produce, notably grinding wheat and other grains to make flour. Mills of various kinds have long been powered by falling water. Writing about human settlements, Griffith Taylor observed, 'Throughout medieval and modern history the advantages of a site near a waterfall have been recognized.'[3] Taylor identifies a type of settlement called 'Fall Towns'. In many cases, these settlements originated because of the falls, but in others development first began for different reasons. In the case of an established settlement, technological innovation might make possible the exploitation of a suitable nearby watercourse as a source of mechanical energy, thus boosting development in the area.

With the increasing role of manufacture, the distribution of energy resources, particularly coal, exerted a much greater influence on the location of industry and consequently of population. In Britain, where the Industrial Revolution can be said to have started, the first locational response to the increased demand for power was the dispersion of mills into hilly areas with rivers and streams whose fast-flowing water could be harnessed to operate machinery. The dispersion of industry from existing towns had other causes, too, including escape from the regulations and fees that generally applied to industrial activities in urban centres, and relatively cheap rural land and labour. The swift-flowing, generally reliable streams of the Pennines in northern England and later in parts of New England were notable among those that offered attractive sites for the new factories that were being built. The replacement of cottage industries and small workshops by large factories was due in part to both the increased use and size of machines, and the limited number of suitable sites along any one watercourse. The diffusion of water-powered industry in Britain became very widespread. Even when coal replaced water power as the source of energy, many of these mills remained on their original streamside sites, their machinery being converted to steam power. This concentration of industrial activity along rivers and streams and the great demand for factory labour encouraged the growth of settlements in many upland valleys. The attraction of rivers as sources of power and water for industry, as well as the topographical influence of steep valley sides on the sites of buildings, roads, canals and railways, tended to impose a linear form on these settlements.

Nodal concentrations of development did occur, however, at major sources of water power, such as Niagara Falls and the Falls of St Anthony. The latter, on the Mississippi River, gave rise to the cities of Minneapolis and St Paul. Another well-documented American example is Spokane, Washington. In the late nineteenth century, the former Indian salmon fishing and trading place at Spokane Falls was transformed into a thriving town as saw mills and flour mills were established there, using

Albert Bierstadt,
The Falls of St Anthony,
c. 1880–87, oil on canvas.

the available water power. Hydro-electric power generation soon followed and before the century was over the city dropped the word 'Falls' from its name.[4]

The Falls of St Anthony, Minneapolis, transformed.

While linear patterns of development occurred along the courses of some rivers, as in the Pennine valleys of Yorkshire and Lancashire, the regional distribution of water power sites also brought about a linear pattern of a different sort. East of the Appalachians in the northeast USA is the Fall Line, a geological boundary where rivers flow from an area of resistant rocks to one where the rocks are more easily eroded. In this boundary zone, stretching in an arc from Alabama to New Jersey, are numerous waterfalls and rapids. Settlements have grown up at many of these sites, sometimes because they mark the head of navigation, the point beyond which large vessels cannot proceed further upstream. Here cargoes were unloaded from the riverboats, the goods being transported further inland by land. At some Fall Line towns, as at many other waterfall settlements, the obstacles to navigation were later removed or bypassed by canals and railways. As we have seen, however, waterfalls, though impediments to transport, can be exploited as sources of power and thus promote industrial development and urban growth. Some of America's most important cities originated in this way as Fall Line settlements, including Washington, DC, Philadelphia and Baltimore.

Among the first of America's industrial towns were those that exploited the water power resources of the rivers of the north-eastern states, mainly for the manufacture of textiles. Early cotton mills established at the existing port towns of Baltimore and Providence were small, contributing little to urban growth. The first large mill, requiring a permanent labour force, was built at Waltham, Massachusetts, a place where falls on the Charles River contributed greatly to the town's early industrial development. The mill owners opened a boarding house there for factory workers, thus increasing the size of the existing settlement. It was water power supplied from the Merrimack River that led to the development of Lowell, Massachusetts. Between 1826 and 1845 the number of mills there grew from the original one to 33, with Lowell's population reaching 30,000. To meet the accommodation needs of many of the workers, as in Waltham, some factory owners built boarding houses, a practice found also in other New England industrial towns like Manchester, New Hampshire and Lawrence, Massachusetts. By 1850 Lowell, with 33,000 residents, was larger than Chicago, Detroit and San Francisco but, in common with other industrial centres of its kind, the town failed to grow into a great city. Nevertheless, the pattern of human settlements in the USA still reflects the influence of water power and its importance in the development of nineteenth-century industry.

In response to some of the worst aspects of nineteenth-century urban development, including slums and pollution, many visionaries and reformers came up with ideas and proposals to improve living conditions in town and country. Perhaps the most influential was Ebenezer Howard, author of *Tomorrow: A Peaceful Path to Real Reform*, published in 1898, and later republished as *Garden Cities of Tomorrow*. The first edition of the book presents the author's Social City concept, illustrated with a theoretical diagram showing a 'Group of Slumless, Smokeless Cities'. Within a circular area 18 kilometres in diameter lie a Central City and a ring of six Garden Cities, arranged within an area where agricultural and forestry uses predominate. Remarkably, in this area there are indicated fourteen waterfalls, each one associated with a reservoir. These artificial features all form part of Howard's ingenious plan for the supply of water and energy to the region. Water collected at lower levels would be pumped to higher-level reservoirs for storage and distribution. In its descent, the water could be harnessed as a source of mechanical or hydroelectric power. The energy needed to pump water to the higher levels would be obtained from 'windmills', either directly, through wind-powered pumps, or indirectly, using electricity generated by the wind.

On the subject of waterfalls, Howard's feelings were both aesthetic and practical. He wrote:

> Unfortunately, under our present immoral and selfish methods, the general weal is little regarded, and the manufacturer, in seeking cheap motive-power, is too apt to deem that cheapest which costs him least, even though it costs society most, and so we hear of waterfalls being desecrated, and as objects of beauty well-nigh destroyed at the advent of some new industry, such as the manufacture of aluminium. But there is no delusion like the delusion of irrational egoism; and if we will but seek the welfare of society, waterfalls, far from being destroyed by manufactures, will be created for them.[5]

Whatever flaws there may be in Howard's scheme for the creation of waterfalls to serve his Social City, his ideas were certainly in advance of his time in the matter of renewable energy, as well as town and country planning.

Among the many new settlements that developed as a result of the industrial exploitation of waterfalls, New Lanark in Scotland is one of the most famous. The Falls of Clyde, near Lanark, had long been known to connoisseurs of landscape beauty. As we

Francis Nicholson,
*Fall of the Clyde at
Stonebyers, c.* 1809–10,
watercolour.

have seen, by this time the falls were a popular tourist attraction. They were also being used to supply power for industry. In 1783 Glasgow businessman David Dale (1739–1806) persuaded Richard Arkwright (1732–1793), inventor of the water-powered yarn-spinning machine, to join him in a partnership to establish a cotton-spinning mill below the Falls of Clyde. The partnership with Arkwright was short-lived, but by 1793 Dale's mill employed 1,300 workers who were fed, clothed and housed under a regime

New Lanark, built
in the Clyde Valley,
southern Scotland.

of benevolent paternalism. Robert Owen bought the New Lanark complex after he married Dale's daughter. Owen's attempt to create there a much more humane living environment than was normal in industrial communities at that time made him and New Lanark world-famous.

Across the Atlantic, another famous place, Niagara Falls, was by now, like the Falls of Clyde, both a popular tourist attraction and a recognized source of mechanical energy. By the mid-eighteenth century, French and British pioneers had established mills near Niagara Falls, using a minute fraction of the potential energy available at the site. Saw mills were among the earliest industrial enterprises to exploit this abundant water power as well as the rich timber resources of the region. The first saw mill at Niagara was built in 1725. Until the 1880s, however, Niagara's mills drew their waterpower from the rapids upstream, not from the cataract.[6]

The beginnings of human settlement at Niagara preceded the exploitation of the falls as a source of energy. Native Americans had long used this locality as a gathering place and trading centre, the rapids and falls of the Niagara River interrupting transport by canoe and necessitating lengthy portage. The importance of Niagara grew when trade with the Europeans began and the British and French built forts and mills to facilitate their exploitation of the country.[7] As this part of North America was opened up by European colonizers, tourism began to make a considerable contribution to the development of the Niagara Falls area.

In the early nineteenth century, the large-scale use of Niagara Falls for mechanical power became evident in the Milling District, situated near the American Falls. By the 1850s, industrial development based on water power had transformed the landscape. Mills sited near the brink of the cataract dominated the skyline. The Niagara Falls Hydraulic Power & Manufacturing Company, first chartered in 1853, built a 1,220-metre-long hydraulic canal to take water from the Niagara River above the falls to mill sites downstream. The construction was completed in 1861, but for over a decade financial problems hampered development. With new ownership in the mid-1870s, various manufacturers

Tailraces from US water-powered mills at Niagara Falls are seen discharging into the river downstream, 1880s, photograph.

were attracted to the mill sites above the American Falls. Initially, each factory on the cliff-top used its own waterwheel and had its own tailrace. The visual impact of this development was considerable, generating much controversy. While the sight of industrial development on the cliff-top and, particularly, the numerous artificial waterfalls formed by the discharge of the tailraces into the Niagara Gorge, inspired admiration among many, others saw the development as a defilement of nature.[8]

By this time electricity had arrived at Niagara Falls, which in the 1870s and '80s was illuminated first by arc lights, then by incandescent lamps. During this period in Europe and North America, increasing use was being made of electricity in individual buildings like factories, railway stations, department stores and residences of the wealthy, mainly for lighting. Although steam engines were normally used to drive the generators, some used water power for the purpose. Soon, water power was being harnessed to provide electricity to groups of buildings and small communities in the USA and elsewhere. One of the earliest central hydroelectric power stations was that built in 1881 by the Niagara Falls Hydraulic Power & Manufacturing Company at its site near the American Falls. This supplied electricity to light the nearby

J. S. Pughe, 'Save
Niagara Falls – From
This', 1906, print.

village of Niagara Falls and provide power to some of the Mill
District factories.[9]

In Europe, too, the potential of hydroelectric power was
being recognized, especially in countries like Switzerland and
Norway, where terrain and limited natural resources constrained

economic development but where there were abundant reserves of untapped water power. In the first decade of the twentieth century, Norway's greatest waterfall, Rjukanfossen, was harnessed to supply the electricity needed for large-scale chemical fertilizer works. This led to the development of the industrial town of Rjukan, with a population of about 9,000 where a mere 50 families had lived before. With technological changes over time, the population has since declined to less than 4,000.

As Lewis Mumford observes, 'In the application of power, electricity effected revolutionary changes: these touched the location and the concentration of industries and the detailed organization of the factory – as well as a multitude of inter-related services and institutions.'[10] While at first technical limitations and high costs restricted the transmission of electricity, later technological advances made it possible to distribute electrical energy over much larger areas efficiently and economically. Hence there is very little incentive to locate industry and other consumers of electricity close to the generating stations, except in cases where power consumption is extremely high, as in the case of aluminium smelting. It was for this reason that the British Aluminium Company established its plant at the Falls of Foyers, Scottish Highlands, in 1896. Since then the availability of large potential hydropower resources has influenced the location of some of the world's largest aluminium smelting plants, including the Alcan works at Kitimat, British Columbia, constructed in the early 1950s. Here a deepwater port for the import of bauxite and the export of aluminium was another important locational factor.

Even as generators of large quantities of power, waterfalls are not usually generators of great cities. While many human settlements are sited beside waterfalls or rapids, something commonly reflected in place names, only in a relatively few cases did this factor alone account for considerable urban growth. One big city that can claim to be largely the creation of a waterfall is Minneapolis which, together with St Paul on the other side of the Mississippi, has a population of over 2.75 million. The harnessed power of the Falls of St Anthony played an important role in urban development, particularly in the early days of the

Rjukanfossen,
Telemark, Norway,
in the 1890s.

settlement when it was an important saw milling and flour milling
centre, but from the 1930s the attraction of the falls as a source of
energy declined as abundant cheap electric power became readily
available all over the country. Possibly more significant as a factor
in the Twin Cities' urban growth was the waterfall's influence as
an obstacle to transportation along North America's greatest
navigable river.[11]

Arguably the world's most famous waterfall cities are the
two municipalities of Niagara Falls, one in the USA, the other in

Industrial plant at
Rjukanfossen.

Canada. Their combined population of over 130,000 is augmented by the millions of tourists who visit the cataract annually. Since the early nineteenth century, tourism has contributed significantly to the growth of the settlements on both sides of the international border. Tourism has been responsible for considerable urban development in many parts of the world, particularly on the coast and in areas of scenic beauty. In this, waterfalls have played an important role as attractions, a notable few as destinations.

Gaspard Dughet, *The Falls of Tivoli*, c. 1661–3, oil on canvas. Since antiquity, Tivoli has been a popular destination for cultural tourists.

11 Waterfalls and Tourism

'Going to see Iguazú Falls? Lots of people go there. If you do, better
stay on the Brazilian side. Only good hotel.'
'Are they worth a visit?'
'Maybe. If you like that sort of thing. Just a lot of water if you ask me.'
Graham Greene, *Travels With my Aunt*[1]

Early tourism, waterfalls and 'the tourist gaze'

Since novelist Graham Greene wrote the above fictional con-
versation, the tourist facilities at Iguassu Falls have developed
considerably on both sides of the Brazil–Argentina border.
Fortunately for the tourist industry, lots of people do enjoy
watching lots of water tumble over rocky cliffs. Since its origins
in antiquity, tourism has been a response to the urge to experience
the wonders of the natural world and some of humanity's great
achievements. Among the natural marvels, waterfalls have long
been popular attractions.

To be a tourist, however, it is not only necessary to have the
desire to enjoy new experiences away from the everyday environment; one must have the leisure and means to do so. In the
worlds of ancient China and Rome, only people from the wealthy,
leisured classes travelled for pleasure, seeking places in which
they could enjoy a variety of novel delights in an escape from
the sometimes irksome routine of normal life. For the wealthy
of ancient Rome, Tivoli provided a convenient retreat, and there,
among hills, woods, streams and waterfalls, the rich and powerful
built pleasure palaces set amid parks and gardens enlivened with
fountains and artificial cascades. Tivoli was also popular with the
wealthy of later centuries. Renaissance aristocrats built magnificent palaces and gardens there, creating many artificial waterfalls
to augment the natural ones in the area. On the other side of the
world, high-class Chinese, particularly those with a scholarly or
artistic bent, were attracted to mountains and cascading streams,
which they found conducive to repose and contemplation. As

discussed earlier, waterfalls were particularly popular as sources of aesthetic pleasure and inspiration.

Iguassu Falls, on the Brazil–Argentina border.

We have seen, too, that, in many cultures, waterfalls were sites associated with the supernatural and often revered as sacred. Sacred sites traditionally attract pilgrims, and some waterfalls became places of pilgrimage, those at Nachi in Japan, at Tirupati and Badrinath in India and Haiti's Saut d'Eau near Villa Bonheur, for example. Historical accounts of tourism commonly refer to religious pilgrimages as an early form of this activity. Others make a distinction between religious travel and tourism, the latter being characterized as travelling for pleasure. If there is truth in Chaucer's account of the Canterbury Pilgrims, however, some of those who went on pilgrimages used the journeys as opportunities for pleasures of an earthly kind as well as for spiritual benefit. The modern tourist may also be regarded as a kind of pilgrim, but, as Donald Horne explains, 'It is no longer God that is sought, but regeneration.'[2] For many, the beauties and wonders of nature, including natural landscapes, are believed to be uplifting and a source of spiritual refreshment. This is one of the reasons why waterfalls have such great appeal, although in the case of more famous falls, especially those of exceptional size, it may be their status as tourist icons that attracts many of the visitors.

As tourist attractions, waterfalls may well have a history as long as that of tourism itself. MacCannell defines a tourist attraction as 'an empirical relationship between *tourist*, a *sight* and a *marker* (a piece of information about a sight)'.[3] For a waterfall, that information could be that the place is sacred, that it is a source of aesthetic pleasure or that it is where certain recreational activities can be enjoyed. It is often a combination of two or more of these. For some tourists, the main attraction may be largely that the waterfall is a famous sight, familiar from numerous images and somewhere to be visited because of the satisfaction achieved by experiencing the original authentic object on the spot. Niagara Falls is the prime example of this. It is, in effect, one of tourism's most sacred sites.

Relatively few tourists today will be motivated to visit a waterfall for the purpose of religious pilgrimage. Probably most of them go in the expectation of aesthetic pleasure, believing that they will experience satisfying emotions at the sight of the water falling from a height in a scenic natural setting. Many go with the intention of enjoying pleasures other than those of a strictly aesthetic kind, although the beauty of the place may play an important role in the choice of the location. Walking, bathing, picnicking, fishing and photography are among the many recreational activities that are commonly enjoyed at waterfalls in addition to the purely aesthetic experience. Waterfalls are examples of what Urry terms 'objects' of 'the tourist gaze'.[4] As landscape features, falls are not only widely regarded as beautiful, but, unlike trees, for example, they are not commonly encountered

Herbert G. Ponting,
Fuji and the Shiraito
waterfall, *c.* 1905,
photograph.

in everyday life. For this reason, they are especially appealing,
delightful curiosities of nature that attract visitors from afar.
Tourists seek experience of this kind in order to help separate
themselves from the often dull round of normal life.

Access can be a problem, however. Waterfalls are commonly
found in places that are often very difficult, even dangerous, to
reach. Typically, they occur in rugged, mountainous areas, very
often hidden in precipitous ravines made hazardous by unstable
slopes and slippery rocks as well as by powerful torrents narrowly

confined between steep cliffs. Approaches to them are commonly made all the more difficult by tumbled boulders and dense woodland. Before the advent of large-scale tourism, the only waterfalls that were likely to be known to people other than those who lived in their vicinity were ones that could be seen or heard from important, well-travelled roads, or which interrupted water transport on major navigable rivers. Among the waterfalls that are mentioned in early British travel literature are those that still can be seen from the old packhorse bridge over the River Ure at Aysgarth, in Wensleydale, and falls that, in quieter times, could be heard from the market town of Kendal on the London to Carlisle road.

Tourist development
at a Japanese waterfall,
c. 1900.

High Force, on the
River Tees in the north
of England, has been a
popular tourist attraction
since the 18th century.

In their descriptions of excursions to scenic places, some
eighteenth- and nineteenth-century travellers dwelt at length on
the difficulties and dangers they suffered in order to visit waterfalls. Arthur Young, an eighteenth-century visitor to High Force,
recorded his tourist experience in his journal. He and his companions had to use their hands and feet to descend, 'almost like a
parrot', then 'crawled from rock to rock and reached from bough
to bough' until they arrived at the foot of the 'noble fall'.[5]

Over two decades later, John Byng, later fifth Viscount Torrington, described his visit to High Force in 1792. He and his party followed their guide 'thro' many hilly, boggy fields . . . endured a most fatiguing descent, and a very dangerous crawl at the river's edge, over great stones, and sometimes up to our knees in water' in order to reach the foot of the waterfall.[6] With the improvement of road transport and the advent of railway travel, more and more people, including members of the growing middle class, went on trips to the country to enjoy beautiful landscapes.

Many of these leisure travellers recorded their experiences and impressions in journals such as those mentioned above and, when published, these books provided useful information for people who wished to follow in their footsteps. By the eighteenth century, books were being written and published specifically to guide travellers in search of landscape beauty and places of interest. Some of them indicated precise spots from which to appreciate views regarded by the connoisseur as being particularly admirable. Among the landscape features most often recommended in such guidebooks were waterfalls.

Authors often added a caveat about excursions to waterfalls. Describing the Lodore Falls in his *Guide to the Lakes* (1784), Thomas West warned, 'It is the misfortune of this celebrated waterfall to fail entirely in the dry season.'[7] Comments of this kind are frequent in guidebooks from West's time to this day, readers commonly being advised that the falls are best seen after rain.[8]

Waterfall access and amenities

The difficulties and dangers that tourists often encountered when trying to visit waterfalls provoked complaints and led to suggestions for improvement. Arthur Young had this to say in his account of a visit to the English Lake District:

> There are a vast many edges and precipices, bold projections of rock, pendent clifts [*sic*] and wild romantic spots, which command the most delicious scenes, but which cannot be reached without the most perilous difficulty: To such points

of view, winding paths should be cut in the rock, and resting places made for the weary traveller . . . At the bottoms of the rocks also, something of the same nature should be executed for the better viewing of the romantic cascades, which might be exhibited with a little art, in a variety that would astonish.[9]

West commended landowner Sir Michael le Fleming for making 'a convenient path' to a cascade in the grounds of Rydal Hall, observing, 'This gentleman's example in opening up a road to the fall recommends itself strongly to others of this country, which abounds with so many objects of curiosity, and which all travellers of the least taste would visit with pleasure, could they do so with convenience and safety.'[10]

Waterfall 'improvements' during the Romantic period were largely associated with improved pedestrian access. No doubt most visitors welcomed these amenities but, as we have seen, others, like Dorothy Wordsworth, were sometimes critical. For her, developments at the Falls of Bruar in Scotland gave particular offence: 'At present nothing can be uglier than the whole chasm of the hillside with its formal walks . . . It does not surely deserve the name of pleasure-path.'[11]

The commercial potential of waterfalls as tourist attractions was soon recognized. Charges were demanded for access to some sites, guide services being offered at others. Ingleton, in a district abounding with waterfalls, was among many Yorkshire villages that experienced a boom in tourism, brought about largely by the arrival of the railway. A Victorian guidebook writer described the difficult terrain that early visitors had to traverse in order to see nearby falls:

meeting with such obstructions of rock, and water, and hanging forest, as well-nigh baffled progress. In some places it was necessary to swing from tree to tree, and spring with utmost caution on to projecting bosses of rock, lest a false step should have launched him into some yawning watery gulf, deep below. Not long after this the register of fatal accidents began.[12]

The influx of tourists gave a strong commercial incentive to improve access from the village to the waterfalls. In 1884–5 the Ingleton Improvement Committee was formed to provide safe and easy access to the falls. Paths were made and bridges constructed to enable visitors to enjoy the river scenery in greater comfort and safety. By the 1890s, it was claimed, 'Happily, now such improvements have been made that the two glens are accessible even to infirm pedestrians – the wielder of the crutch may safely venture – and the scenery of them both, which involves a walk of some four or five miles, viewed with ease and composure in the course of a summer afternoon.'[13] In the 1930s the Ingleton Advertising Association emphasized the safety of the walk in a detailed description that mentions its series of steps, platforms and bridges, wayside seats and places where refreshments could be purchased.[14] At first, when the two properties were in separate ownership, an admission fee was charged at the entrance of each of the valleys. This situation encouraged aggressive commercial competition, resulting in a rash of outdoor advertisements that disfigured the approaches to the beauty spots. Things improved after both attractions fell into common ownership, although the poor quality of signage along the footpath gave offence to some.[15] The Ingleton Waterfalls Walk remains one of the popular tourist attractions of the Yorkshire Dales National Park.

For some tourism-promoters wishing to fully exploit the commercial potential of waterfalls, the improvement of pedestrian access was not always enough. Nineteenth-century technological advances made possible various modes of transport that conveyed tourists to remote and inaccessible places, including mountaintops and waterfalls. Already popular as a tourist attraction, Switzerland's Reichenbach Falls received a boost in popularity with the construction of a funicular railway there in 1899, cables coming increasingly into use for tourist transport in the following century. A modern cable car provides easy access to the top of Montmorency Falls near Quebec, Canada. Another, at Wulai in Taiwan, takes visitors up to the top of the dramatic waterfall where, in addition to the scenic attractions, an amusement park

Kayaks at the foot of a waterfall in Milford Sound, New Zealand.

has been created to diversify the tourist experience. For access to the foot of the Wulai Waterfall, the visitor has a choice between a twenty-minute walk or a five-minute ride on a miniature railway.

Some waterfalls can be reached by boat. Best known, perhaps, is Niagara's *Maid of the Mists*, a boat trip that originated in 1846. Around the world, a wide variety of vessels are used for the purpose, some catering for the more adventurous tourist. In New Zealand a jetboat takes visitors to the foot of Huka Falls, the passengers experiencing speeds of up to 80 kilometres an hour and a 360-degree spin in which the boat turns on its own length. Also in New Zealand, some waterfalls are best seen from boats that cruise the fjords of South Island's spectacular west coast. Norway offers similar tourist experiences along its famous fjord coastline. Many waterfalls are now commonly visited by air, usually in light planes or helicopters. Australia's Jim Jim and Twin Falls, both in Kakadu National Park, are often difficult to reach by dirt road during the wet season when they are at their scenic best, but can be viewed from the air on tourist flights. Even more remote are Venezuela's Angel Falls (see page 65), but tourists now go by plane to see this famous but formerly little-visited natural wonder, enjoying spectacular aerial views of the world's tallest waterfall. At Kaieteur Falls, Guyana (see page 40), there is even a small airstrip enabling tourists to reach this remote spot by flying from the capital, Georgetown, a trip of little over an hour in each direction. The town of Victoria Falls is served by an international airport which caters for large passenger aircraft, but tourist flights over the celebrated cataract are made in light planes that can fly low over the chasm into which the Zambezi tumbles. While it is an exciting experience to view waterfalls from the air in this way, for visitors on the ground tourist aircraft can be noisy intrusions that mar the enjoyment of nature's splendour. They contribute to the visual and noise pollution that are unfortunate concomitants of improved access to waterfalls.

Tourists on a cruise ship enjoy a close-up view of one of the many waterfalls in Milford Sound.

Waterfall destinations and attractions: the tourist experience

While many waterfalls in various parts of the world are exploited as tourist attractions, a very few may be regarded as destinations in their own right. In the tourist industry, attractions are features that contribute to the appeal of a tourist destination, which may be a country, region, city or national park, for example. Perhaps only two or three of the world's waterfalls can be classed as tourist destinations. It certainly may be argued that Niagara, Victoria and perhaps Iguassu Falls can each be classed as a destination. A destination has been defined as 'a place having characteristics

Helicopter services
enable tourists to
enjoy spectacular
aerial views of some
of the world's great
waterfalls, including
Sutherland Falls,
inland of Milford
Sound.

known to a sufficient number of potential visitors to justify its consideration as an entity, attracting travel to itself, independent of the attractions of other locations.'[16] Visitors to these places often spend several days there or even longer, instead of the shorter periods – possibly only an hour or so – normally allowed for individual attractions. On tourist air excursions to Kaieteur Falls, one of the world's most magnificent cataracts, visits usually last only a couple of hours.

In the early days of tourism at Niagara Falls, visits typically lasted several days, even weeks. The development of Niagara Falls for mass tourism began with major improvements in transport. As early as the 1830s, with access by canal and rail, the setting of Niagara Falls was being transformed by uncontrolled tourism development. Private developers acquired the best viewpoints, then imposed charges on visitors who wished to gain access in order to enjoy the spectacle. By 1860, fences and gatehouses surrounded the Falls on every side. Apart from the sight of the cataract itself, both the American and the Canadian Falls, there were other attractions for the tourist to experience. These included natural landscape features, such as the gorge, rapids, whirlpool and the Cave of the Winds, and 'artificial' attractions such as the *Maid of the Mist* boat ride and the professional photographic service on Luana Point.

According to the English traveller Isabella Bird (1831–1904), at this time many in her home country supposed America to be 'a vast tract of country containing one town – New York; and one astonishing natural phenomenon, called Niagara'.[17] These were the two sights that had to be seen by any tourist who made the journey across the Atlantic. Bird's record of her visit to Niagara Falls in 1854 provides a vivid account of the tourist experience of Niagara in the mid-nineteenth century. On arrival at her hotel, Clifton House, on the Canadian side of the river, Isabella went straight to the cliff edge to have her first view of Niagara Falls. While immediately impressed by the sight of the great cataract, she was at the same time struck by the disfigurement caused by industrial development beside the American Falls and tourism development on the Canadian side. 'A whole

collection of mills disfigures this romantic spot', she wrote. 'And even on the British side, where one would have hoped for a better state of things, there is a great fungus growth of museums, curiosity-shops, taverns, and pagodas with shining tin cupolas.'[18] It was not long before she was accosted by a man offering her a carriage tour of the sights at the cost of the hotel, and it was with some disappointment that she returned to her hotel 'enduring a whole volley of requests from the half-tipsy drosky-drivers who thronged the doorway'. Sightseeing expeditions from the hotel could be quite vexatious. The visitor,

Early tourism at Niagara Falls, 1870s, lithograph.

> yielding to the demands of a supposed necessity, is dragged a weary round – he must see the Falls from the front, from above, and from below; he must go behind them, and be drenched by them; he must descend spiral staircases at the risk of his limbs, and cross ferries at that of his life; he must visit Bloody Run, the Burning Springs, and Indian curiosity shops, which have nothing to do with them at all.[19]

Only after the almost compulsory ritual of 'doing Niagara' could the tourist with a sensitive appreciation of landscape steal away quietly to gaze at the cataract. Getting about the area involved dealing with vociferously competing carriage drivers, and paying fares, fees and tolls for the use of vehicles, roads, bridges and access to viewpoints, as well as expenditure on refreshments, souvenirs and the like. Along the way were 'tea-gardens, curiosity-shops, and monster hotels, with domes of shining tin', but despite the quarrelsome drivers, the 'incongruous mills, and the thousand trumperies of the place', Isabella Bird was able to enjoy 'the perfect beauty of the scene . . . the joyous realisation of [her] ideas of Niagara. Beauty and terror here formed a perfect combination'.[20] Very different were the circumstances when, dressed in 'an oiled calico hood, a garment like a carter's frock, a pair of blue worsted stockings, and a pair of India-rubber shoes', the intrepid English-woman, accompanied by a 'negro guide', walked behind the Horseshoe Falls to Termination Rock. It was for her a terrifying experience, one 'far better omitted' by tourists, especially anyone 'who has not a very strong head'. Far more enjoyable for her was the ascent of an observation tower from which she had 'a very good bird's-eye view of the Falls, the Rapids, and the general aspect of the country'. Clearly, not all tourist facilities at Niagara were objectionable to her.[21]

Later in her travels, Bird visited Quebec City, from which she drove out to see the nearby Montmorency Falls. Of this scenic spot she wrote, ' Montmorenci [*sic*] gave me greater sensations of pleasure than Niagara. There are no mills, museums, guides, or curiosity-shops. Whatever there is of beauty bears the fair impress of its Creator's hand.'[22] Today, Montmorency Falls no longer remain in this pristine state. Apart from the 83-metre waterfall ('30 metres more than Niagara') and an historic house, now a restaurant, bar and reception centre, the Montmorency Falls Park boasts many other attractions, including a cable-car ride, panoramic stairways, two suspension bridges over the Falls and the rift, four belvederes with panoramic views, a cliff-top path, snack bars, three boutiques, an interpretation centre, a tourist information office, archaeological and historical displays, and in winter the

Sugarloaf (a hill of ice formed by mist from the Falls) and ice-climbing, with courses for beginners. Special events there include firework displays, theatrical productions and exhibitions.

The scale of development at Montmorency Falls remains far less than that at Niagara Falls. Currently listed Niagara attractions include the Daredevil Museum of Niagara, the Wintergarden, an Aquarium of Niagara, the Schoellkopf Geological Museum, the Summit Park Mall, a Factory Outlet Mall, an Artisans Alley, the Hyde Park Golf Course, the Lavinia Porter Manse, the Seneca Niagara Casino, the Niagara Falls Convention and Visitors' Bureau, the Niagara Spanish Aero Car, Skylon Tower, Marineland and the Niagara Parks Butterfly Conservatory. Clifton Hill, alone, has much to offer, such as Falls Tower, Guinness World of Records, Dinosaur Park Miniature Golf, Ripley's Moving Theater, a Fun House, a Mystery Maze and the House of Frankenstein. The list continues, including old favourites such as *Maid of the Mist* and the Cave of the Winds Tour, although the cave itself was deliberately destroyed for safety reasons after a massive rock fall in 1954. The Falls of Niagara have been dramatically illuminated at night since 1925.

The flow of water at Niagara Falls has been greatly reduced by diversion for the generation of hydroelectricity but the cataract remains a spectacular sight. Elsewhere, power schemes have largely destroyed waterfalls that had previously been notable tourist attractions, Australia's Barron Falls and Jamaica's Roaring River Falls, for example. At some waterfalls reduced by power schemes, the flow of water is turned on periodically for the benefit of tourists. Some have an advertised schedule of times when water is released over the falls for the enjoyment of visitors. Among these limited appearance waterfalls are Maria Cristina Falls in the Philippines, Sweden's Trollhatten Falls and Italy's Marmore Falls.

At Africa's great waterfall destination, Victoria Falls, tourism came much later than at Terni or Niagara. 'Discovered' by Livingstone in 1855, the year after Isabella Bird's North American tour, tourism development there had to await the arrival of the railway in 1904, when today's luxurious Victoria Falls Hotel

The Aratiatia Rapids on the Waikato River, New Zealand, before a scheduled release of water from the hydro-electric dam upstream.

had its humble beginning. Nowadays, most tourists arrive at Victoria Falls by air, using the international airport just outside the resort town with its population of over 30,000. By the end of the twentieth century, the monthly average number of tourists was a similar figure. While most visitors come mainly to see one of the world's greatest natural spectacles, the cataract on the Zambezi, other local attractions are also exploited by the tourist

The Aratatia Rapids
after the opening of
the dam floodgates.
A nearby noticeboard
informs visitors of
the times when this
spectacle can be seen.

industry, notably African culture and wildlife. As at Niagara Falls, many additional attractions have been created at Victoria Falls, including golf courses and casinos, museums and souvenir shops. A plethora of roadside hoardings and signs advertise a wide range of goods and services ranging from wildlife safaris and African cultural performances to bungee jumping and fast-food outlets. Victoria Falls boasts one of the world's highest commercial bungee jumps, operated from the road and rail bridge on the Zambia–Zimbabwe border below the Falls. White-water rafting on the rapids downstream is another popular extreme sport that attracts many of the younger visitors. More tranquil boat rides can be enjoyed above the Falls, while joy flights by light plane or helicopter afford spectacular views of the cataract from the air.

Development in the immediate environs of Victoria Falls has been better controlled than that at Niagara. Both sides of the African cataract are protected in national parks, a form of landscape conservation area that did not exist when the commercial exploitation of Niagara Falls began. Indeed, public outrage at what seemed to many nineteenth-century visitors to be the desecration of Niagara Falls was one of the main factors in the rise of the national park movement in the USA. Another was the promotion of tourism, largely supported by the railroad companies. Waterfalls contribute greatly to the scenic values of many areas that were declared national parks in America and elsewhere. Among the principal scenic attractions of two of the world's first and most famous national parks, Yellowstone and Yosemite, are waterfalls of exceptional grandeur, and falls of all kinds are important landscape features in many other scenic reserves all over the world. These range from the great cataracts of Victoria, Iguassu and Kaieteur to the many falls and cascades that are to be found in Australia's Lamington National Park, reputed to contain 500 waterfalls, and the Yorkshire Dales, where many of England's loveliest and best-known falls are to be seen.

Visitor activities at waterfall sites

Even in protected areas such as national parks, where waterfalls are developed for public enjoyment, the provision of safe and convenient access may create visual intrusions like paths, steps, viewing platforms, fences and signs that can detract from the aesthetic experience. As we have seen, some popular waterfalls can be reached by means of a variety of forms of transport and for many people the novel rides are part of the enjoyment of the visit. For some visitors, however, 'artificial' attractions like scenic rides detract from the experience of 'unspoiled nature' that they seek. The problem may be further exacerbated when visitors engage in activities that interfere with the enjoyment of others. The mere presence of other people, especially when they come in large numbers, can spoil the pleasure for some, and the easier the access the greater the likelihood of crowds of visitors. Bathing and picnicking have long been pleasures associated with water-falls. At some falls, too, good fishing can be had, making them

A Victorian picnic at Thomassin Foss, near Goathland, North Yorkshire.

Rock climbers at a waterfall in the Lake District, England.

popular spots for anglers, and, in common with all beauty spots, waterfalls attract photographers. Other outdoor activities commonly enjoyed at waterfall sites can be much more intrusive. Many waterfalls are popular with rock climbers and abseilers who practice their skills on cliffs beside the falling water, some even ascending or descending through the falls themselves. Canyoning has become a popular sport that often involves the descent of waterfalls. In some parts of the world, prolonged sub-zero temperatures create conditions for ice climbing at waterfall sites. Not all waterfall climbs are challenges for the expert, and in Jamaica, Dunns River Falls are climbed by thousands of tourists every year, making this activity one of the island's most famous attractions (se p. 40). Various forms of boating are enjoyed at some falls. For the adventurous, white-water rafting, sledging and canoeing are extreme sports that are commonly available in the vicinity of major waterfalls, while some courses actually include low falls. New Zealand's North Island boasts what is claimed to be the world's highest commercially rafted and sledged waterfall, Tutea Falls, a seven-metre drop on the Kaituna River where it descends over a series of rapids and falls together known as Okere Falls.

The commercialization of a waterfall attraction is nowhere more dramatically demonstrated than at Dunns River Falls, on Jamaica's north coast. Easily accessible, just outside the resort town of Ocho Rios, and crossed by the main road between Kingston and Montego Bay, this beautiful cascade has the additional attraction of tumbling onto a white sand beach ideal for sea bathing. For many years the natural tiered formation of the falls has tempted visitors to climb them. For half a century, this activity has been promoted by the Jamaican tourist industry, making Dunns River Falls, with its famous climb, one of the country's most important attractions. It receives about a million visitors annually. The landscaped setting, crafts market, food and refreshment outlets and other amenities, as well as the crowds both beside and in the cascading river, appeal more to gregarious fun-loving tourists than to those who prefer the peaceful enjoyment of natural landscape beauty. For this reason, there are many visitors to Jamaica who now seek their enjoyment of waterfalls elsewhere on the island. This, together with local recognition of the tourism potential of waterfalls as demonstrated clearly by the commercial success of Dunns River, has encouraged development at several other Jamaican falls.[23]

A brochure advertising white-water rafting on the Kaituna River in North Island, New Zealand.

Safety issues

It is not only the natural environment that can suffer damage in consequence of tourism. As noted earlier, tourists who enjoy outdoor recreation in areas of landscape beauty also run the risk of injury, even death, when they visit wild places like waterfalls.

This problem and the tourist industry response has already been described in connection with the development of the Ingleton waterfalls. Despite precautions, including physical barriers and warning signs, accidents continue to occur at waterfall and ravine sites. Often this is the result of visitor carelessness or foolhardiness, but sometimes misfortune occurs for other reasons. Fortunately tragedies, such as the collapse of a viewing platform at Cave Creek, New Zealand in 1995 with the loss of fourteen lives, are very rare. It is the vagary of the weather and stream flow that continues to take its toll at places such these. Usually, the occasional victim of drowning or falling receives no more than local attention in the media, but sometimes the tragedy is on a sufficient scale to hit the world's headlines. The Swiss canyoning disaster of 1999 which claimed eighteen lives, and the deaths of at least 37 visitors swept away at Thailand's Sairung and Prai Sawan waterfalls in 2007, made stories that captured media attention around the world. In response to these events, steps have been taken to establish warning systems intended to reduce the dangers to those who visit and engage in various recreational pursuits at waterfalls.

Ignoring warning signs, a woman dives from the brink of a waterfall at YS Falls, Jamaica.

Even today, there are many waterfalls that tourists find difficult to reach and that may present hardships and dangers to the unprepared visitor. This is especially true in areas where tourism is in an early stage of development. In September 2008 Australian newspapers reported a story about a Queensland man on holiday in the Khammuan province of Laos who decided to visit the Tadsanam waterfall. This is a mere three-kilometre walk from the tourist's backpacker hotel, where he was advised that that no guide was necessary for the excursion. Setting out along the jungle footpath, the tourist soon lost his way, possibly because rising water may have covered the track. There followed an eleven-day ordeal that nearly cost the man his life in the densely forested, rocky terrain. A helicopter search eventually found him lying beside a waterfall – not the one he had set out to see, but one that nevertheless probably saved his life as a source of drinking-water.[24]

As well as involuntary deaths at waterfalls, many suicides occur at these beautiful and dangerous places. As noted earlier, there is something about waterfalls that the human mind associates with death and the hereafter. For visitors intent on ending their lives at a waterfall there is little that safe paths, fences and warning signs can do to prevent self-destruction. Happily for most visitors it is a love of life and the enjoyment of nature that brings them to these scenic spots.

Waterfalls and tourism promotion

The appeal of waterfalls and their potential value as tourism attractions is now widely recognized. In many parts of the world, including those commonly referred to as 'Developing' or 'Third World', countries have begun to exploit these landscape resources. Some, such as Venezuela, Guyana and Zambia, have the advantage of possessing waterfalls that are already world-famous, while others, like Ghana, Malaysia, Brunei and Vanuatu, publicize falls that are little known globally, attempting to diversify their 'tourism product' by adding to their range of advertised attractions. Commonly, waterfalls are featured in advertisements not as specific attractions in themselves, named falls that tourists are encouraged to visit, but as elements in an idyllic tropical land-scape that is supposed to appeal to tourists in search of a vacation paradise. To these ends, waterfalls are featured in advertisements of all kinds.

One kind of tourism promotion that goes back over a century is the pictorial postage stamp. The first postage stamps to feature waterfalls were a Belgian Congo issue of 1894 and a Salvadorean

Wli Falls, Ghana.

one of 1896. Neither of these countries could be regarded as a pioneer of tourism, but elsewhere images of waterfalls on stamps were undoubtedly used to promote the tourism industry. Jamaica's Dunns River Falls, discussed above, feature on two different issues but the first of that country's waterfall postage stamps shows the now degraded and almost forgotten Llandovery Falls. Issued in 1900, this stamp has an image derived from one of the photographs taken by James Johnston as part of his Jamaican tourism promotion campaign. On the other side of the world, in Tasmania, another photographer and campaigner for tourism, J. W. Beattie, had two of his waterfall images reproduced on postage stamps issued in 1899. One features Russell Falls, still one of Tasmania's most famous scenic attractions. Since that

Tourists at Wli Falls. The man in the middle of the group is using binoculars to observe bats, which have colonized the cliffs above.

Tourist centre,
Wli Falls, Ghana.

time, hundreds of waterfall postage stamps have been issued by many different countries, from Argentina to Zimbabwe.

The tourist activity generated by the attraction of waterfalls now threatens to destroy the resources on which this aspect of tourism depends. For those who visit waterfalls to experience the natural environment and enjoy the scenic beauty, excessive development is an anathema. Highly developed and commercialized beauty spots can loose their attraction, encouraging tourists to go elsewhere for the satisfactions they seek. This, in turn, contributes to the spread of what may be termed tourism blight. Tourism and hydropower generation are major threats to waterfalls and many lovely falls have already been lost or spoiled by these demands.

One of the many waterfalls to be seen in Fiordland National Park, New Zealand.

12 Waterfalls Lost, Despoiled and Threatened

'It is, indeed, the rapidly increasing spoliation, amounting in some cases almost to entire destruction, of so many of the notable waterfalls that has prompted me to the task of compiling this book and to record some descriptions of the most famous of them while they still retain something of their pristine grandeur.'
Edward C. Rashleigh, *Among the Waterfalls of the World* [1]

Rashleigh's book *Among the Waterfalls of the World* remains a classic in the field, gaining importance over the years with the continuing sad loss of many falls. These have often been sacrificed for power generation and water supply purposes to serve industry, agriculture and urban development. Other falls have been exploited commercially as tourist attractions. Waterfalls are more than just attractive features of the landscape that are of great cultural significance; they are valuable economic resources.

In an era characterized by unprecedented levels of species extinction due largely to the exploitation of the earth's natural resources, some distinctive landforms, too, are endangered. With global reserves of coal and oil rapidly vanishing at a time of increasing energy demands, and with water supplies failing to meet growing domestic, agricultural and industrial needs practically everywhere, the disappearance of the world's finest pristine waterfalls looms as a distinct possibility. The strongest protection that can be given to waterfalls is their inclusion in national parks or similar landscape conservation areas. Among the outstanding scenic wonders of America's and the world's first national park, Yellowstone, established in 1872, are magnificent waterfalls. The many falls in the Yosemite Valley, including some of the world's highest, were given protection even earlier when the area was declared a state park in 1864. Yosemite became a national park in 1890. Among other great waterfalls of the world now protected within national parks are Victoria, Iguassu and Kaieteur Falls. Many other national parks and similar conservation areas protect landscapes in which waterfalls are important features, among them New Zealand's Fiordland, Iceland's Jökulsárgljúfur and England's Yorkshire Dales. Within Fiordland are many fine waterfalls, including Sutherland Falls, one of the world's highest, while Jökulsárgljúfur has Europe's greatest waterfall, Dettifoss. The numerous little waterfalls of the Yorkshire Dales, though small by world standards, are charming features of the Pennine landscape and are popular with tourists. With wise management, waterfalls, like other types of attractive scenery, can be developed for tourism in a sustainable way.

Popularity is often one of the most serious problems that threaten waterfalls in areas where the tourist industry exploits these scenic resources. Nowhere is this better exemplified than at Jamaica's Dunns River Falls, which now attracts adverse comment in the tourist literature. Even without commercialization, for many, the mere presence of large crowds of people can detract from the enjoyment of a visit to a waterfall. Other problems associated with large numbers of visitors include the erosion of footpaths and surrounding areas, damage to plants, disturbance of wildlife, litter, graffiti, noise and other forms of pollution.

Waterfalls dry; waterfalls drowned

While the effects of tourism can detract from the aesthetic enjoyment of a waterfall, the upstream diversion of water for power generation, irrigation or other purposes inevitably results in the diminution or complete loss of falls where the scenic beauty often depends largely on a good flow over the brink. This is not just a recent problem. Describing a visit to Dyserth in North Wales, eighteenth-century traveller Thomas Pennant wrote, 'A waterfall in the deep and rounded hollow of a rock, finely darkened with ivy, once gave additional beauty to this spot; but of late the diverting of waters to a mill, has robbed the place of this elegant variation.'[2]

Since then, the loss or diminution of waterfalls by diversion has multiplied greatly. Among the great waterfalls that have been lost in this way is Rjukanfossen, formerly considered to be one of the most beautiful cataracts in the world, which was sacrificed

The Itaipu Dam, on the Brazil–Paraguay border, has swallowed the Guaíra Falls.

for Norway's industrial development at the beginning of the twentieth century. Many others have suffered substantial reductions in flow, notably Niagara Falls, where the amount of water leaping over the cliffs is less than half of the volume that formerly distinguished the great cataract. An early casualty in this scenic spoliation process was a Scottish waterfall well known to Romantic tourists of the eighteenth and nineteenth centuries: the Falls of Foyers. Edward Rashleigh bemoaned their fate: 'But, alas, in 1895, in spite of loud protests from many famous men and lovers of natural scenery, the falls were handed over to the tender mercies of an Aluminium Company and are today so depleted as scarcely any longer to be worth seeing.'[3] During the late twentieth century, among the great cataracts lost in this way were two in South America: Brazil's Paulo Afonso Falls, now dry because of water diversion upstream, and the Guaíra Falls, between Brazil and Paraguay, submerged by a reservoir formed by dam construction downstream.

At waterfall sites where stream flow has ceased due to upstream diversion, the landscape may retain much of its original beauty, especially where there are spectacular cliffs and, perhaps, forested gorges. Barron Falls and Tully Falls in North Queensland have both been sacrificed for power generation but the tropical gorges there are still dramatically beautiful. In a few cases, where the discharge of water is enormous, the permanent diversion of some of the flow may have a relatively slight effect on the appearance of the falls. Under a treaty signed in 1950 by Canada and the USA, Niagara Falls are protected by the maintenance of a minimum

Victoria Falls. This part of the cataract, known as the Main Falls, lies within Zimbabwe's Victoria Falls National Park.

flow which preserves much of the cataract's scenic beauty while allowing a half to three-quarters of the river's flow to be diverted for the generation of electricity. Even the relatively undeveloped Victoria Falls on the Zambezi River lose some water to hydroelectricity generation on the Zambian side, but the power scheme remains small and the scenic impact is minimal. In the case of small waterfalls, however, diversion of flow inevitably incurs the loss of landscape quality and in some cases once famous beauty spots cease to attract tourists. For example, Jamaica's Roaring River Falls were once the island's major scenic attraction, but today they are practically unknown and unvisited. Their former beauty was sacrificed for a small hydropower scheme completed shortly after the Second World War. It was only after the depletion of Roaring River Falls that the nearby Dunns River Falls rose to fame as one of Jamaica's most popular tourist attractions.

Waterfalls today and tomorrow

While many lovers of landscape beauty are saddened by the loss of waterfalls in the name of progress, the benefits that these developments have brought through the provision of electricity and water supplies are enormous. Travel writer Pamela Watson expresses this vividly in her book *Esprit de Battuta: Alone Across Africa on a Bicycle* (1999). While travelling through the Republic of Guinea, Watson saw 'the remnants of the waterfall' known as the Chutes du Tinkisoo. 'Then it had seemed a shame', but when she entered the town of Dabola, which was now supplied with electricity generated by the harnessed falls, the author was struck by the energy and liveliness of the place in contrast to other places which she had visited in the country. 'Dabola . . . was a sizeable town, as big as Koundara, but with electricity, and what a difference that made.'[4]

The relative cheapness and cleanness of hydroelectricity make it very attractive in countries with hydropower resources that can be readily exploited, especially where alternative sources of energy are scarce and expensive. When considering waterfalls as economic resources, there are cases where it may be difficult

to make the choice between harnessing the falls for electricity or exploiting their scenic beauty for tourism. This decision had to be made by the Jamaican Government in 1989 when there was a proposal to harness the lovely YS Falls for a hydropower scheme. The argument for preservation of the falls prevailed, largely on the grounds of their value as a tourism attraction in a country where the tourist industry is of vital importance to the national economy.

Tourists at YS Falls, Jamaica. These two visitors risk injury, even death, as they leap down the waterfall.

More recently, one of equatorial Africa's most beautiful cataracts, Kongou Falls in Gabon's Ivindo National Park, has come under threat from a planned hydroelectric power scheme associated with a Chinese sponsored iron ore mining project. If carried out, this controversial proposal would not only result in the loss of yet another of the world's spectacular waterfalls but could threaten the declassification of Ivindo National Park as a protected area. Gabon's thirteen national parks were established in 2002 largely to encourage ecotourism as a major sector in the country's future economy.

With growing demands for energy and the continuing expansion of the tourist industry, pressures on waterfalls as resources for economic development are likely to increase. Victoria Falls is a case in point. In recent years, UNESCO's World Heritage Committee has expressed concerns about environmental degradation there:

> Major present problems are the haphazard proliferation of tourist infrastructure and pollution, invasive species and upstream water abstraction. Visual and aural pollution from development are intensifying. There are 20 helicopter and light aircraft flights a day over the falls, disturbing the wildlife, and 40 cruise boats, some being jet boats, ply above them in Zimbabwe. There are bunji-jumping [*sic*], a gorge swing and plans for a balloon to be tethered over the falls, with service buildings on the ground ... several hotels have been built within the site on both sides of the river. And the Zambian President has approved a large 5-star hotel/ convention hall/golf course/marina development with a secondary 4-star hotel and luxury villas planned along the north bank mostly within the World Heritage site. The management of the Parks is not very effective in the face of these urgent pressures for large commercial developments and the presence of a growing local population.[5]

In the case of tourism, preservation as well as exploitation is essential for sustainable use, but for the generation of electricity waterfalls must lose flow, even disappear completely, for their energy to be tapped. Where alternative sources of energy are available it is easier to argue for the preservation of waterfalls on aesthetic grounds, the latter case often strengthened by their value as tourist attractions. One of the advantages of hydropower is that it is a renewable resource, one of several on which we must have to rely increasingly as fossil fuel reserves are exhausted or become less attractive for environmental reasons. If there are great advances in the use of other types of renewable energy resources such as solar and wind power and that derived from sea waves, tidal currents and the conversion of biomass, pressures to harness waterfalls for the generation of electricity may ease, however. It may then be possible to halt the loss of this disappearing landform and, possibly, restore some of those that have been sacrificed for power generation.

Since Rashleigh's time, problems of a different kind have emerged or have grown in significance to threaten the beauty and environmental qualities of waterfalls. The water that Rashleigh saw tumbling over Niagara Falls was already very polluted by waste products from the densely populated and highly industrialized Great Lakes region. Despite recent improvements in river water quality in some areas, water pollution continues to spread, even affecting falls that once seemed safe from environmental degradation. The expansion of tourism has both helped to preserve waterfalls and added to the problems of their development. Other land use changes, too, especially deforestation, have had detrimental effects on waterfalls, including increased turbidity related to soil erosion and greater variation in stream flow, often resulting in dry watercourses for much of the year. Climate change is another important factor affecting river regimes, one that has already had serious consequences in some parts of the world, including the Caribbean. In Jamaica, declining rainfall was noted by historian Edward Long over two centuries ago, a phenomenon then largely attributed to deforestation and related climate change. Today, other human influences, too, are widely believed to be affecting global climate, including rainfall patterns. Whatever the cause or causes, there appears to be little doubt that during the twentieth century many of Jamaica's formerly perennial rivers began to run dry for part of the year.[6]

Malaysia is one country where the consequences of this type of landscape degradation on tourism have been recognized. In a recent internet article published under the heading 'Drying water falls bad omen for Penang's tourism?', Melissa Darlyne Chow warns that two waterfalls in the state of Penang have dried up while three others 'are on the verge of suffering the same fate'. Illegal logging and illegal water tapping by farmers are blamed, and a Waterfall Restoration Committee has been

Khone Falls, Laos. This series of falls and rapids is threatened by dam construction projects planned in several countries of the Mekong Basin.

established by Penang's tourism and environment authority to address the problem.[7]

The fate of the world's waterfalls is inextricably linked with many of the environmental and economic issues that concern us today. These include climate change, renewable energy, increasing demand for water, environmental degradation, conservation and ecotourism. Having played a significant role in the cultural and economic life of humankind since ancient times, waterfalls are now under severe threat from many different sources. As resources for renewable energy and sustainable tourism they are, nevertheless, exhaustible. Globally, waterfalls are in limited stock and are an endangered landform. In terms of the natural landscape, the continued reduction and loss of the world's waterfalls would lead to aesthetic impoverishment that would deprive generations to come of some of nature's most beautiful and sublime sights.

NOTABLE WATERFALLS

Angel Falls (Kerepakupai Vená), Venezuela, is the world's tallest waterfall at 3,212 ft (979 m).

Augrabies Falls, South Africa, is a series of huge cataracts on the Orange River where it tumbles through a gorge cut through bare granite in an arid setting.

Bambarakanda Falls is Sri Lanka's tallest waterfall at 863 ft (263 m).

Barron Falls, Queensland, formerly Australia's grandest cataract, is now much reduced by a hydropower development.

Bridalveil Fall, in California's Yosemite Valley, drops a sheer 620 ft (189 m).

Cascata delle Marmore, Italy, was created by a Roman engineering scheme. With a total drop of 541 ft (165 m), it is the world's tallest artificial waterfall.

Detian / Ban Gioc Falls, are on the Quy Xuan River that forms the border between China and Vietnam.

Dettifoss, Iceland, is Europe's largest waterfall in terms of discharge volume. It falls 144 ft (44 m) over a width of 440 ft (134 m).

Dry Falls in Washington is a 3.5 mile (5.65 km) long line of cliffs some 400 ft (122 m) high, formed by an enormous Ice Age cataract.

Dunns River Falls, Jamaica, cascades onto a popular Caribbean beach. It is climbed by thousands of tourists annually.

Eas a' Chual Aliunn, Sutherland, in the far north of Scotland, descends 660 ft (200 m), making it the tallest waterfall in the British Isles.

Gaping Ghyll Waterfall, in the Yorkshire Dales, drops vertically 344 ft (105 m) into a large cave. It is the tallest unbroken waterfall in Britain.

Gavarnie Falls (Grande Cascade de Gavarnie), descends 1384 ft (422 m) into a glacial cirque in the French Pyrenees. It is commonly regarded as the tallest waterfall in France.

Gocta Falls, Peru, drops 2,530 ft (771 m) in two leaps, making it one of the world's tallest waterfalls.

Goðafoss, Iceland, derives its name from the arrival of Christianity, when figures of old Norse gods were tossed into the cataract. It is 40 ft (12 m) tall and 98 ft (30 m) wide.

Guira Falls, Brazil, formerly one of the world's greatest cataracts, was submerged by a reservoir created when a dam was constructed downstream in 1982.

Hannoki Falls is the tallest waterfall in Japan, at 1,600 ft (500 m).

High Force on the River Tees drops 70 ft (21 m), and is one of the most impressive waterfalls in England (it is sometimes wrongly described as one of the tallest).

Huangguoshu Waterfall, also known as the Yellow Fruit Tree Waterfall, is one of the largest in China and East Asia.

Iguassu / Iguazu / Iguaçu Falls is a huge cataract on the Brazil–Argentina border. It comprises about 275 separate falls 200–269 ft (60–82 m) tall, with a total width of 1.7 miles (2.7 km).

Jog Falls (Gersoppa Falls), India, with a drop of 829 ft (253 m), is a spectacular sight during the monsoon season. Now harnessed for hydropower generation, the waterfall is much reduced during most of the year.

Kaieteur Falls is a magnificent high-volume waterfall on the Potaro River, Guyana. With a drop of 741 ft (226 m), it is about five times the height of Niagara Falls and twice that of Victoria Falls.

Khone Falls, a series of cataracts in southern Laos, near the Cambodian border, are a major impediment to navigation on the Mekong River.

Lodore Falls, in the English Lake District, is a small cascade made famous by Robert Southey, who wrote the onomatopoeic poem *The Cataract of Lodore* in 1820.

Murchison Falls (Kabarega Falls), Uganda, is where the Victoria Nile descends 141 ft (43 m), seething through a rocky channel less than 30 ft (10 m) wide.

Nachi Falls, Japan, with a single vertical drop of 436 ft (133 m), is a revered sacred site.

National 9/11 Memorial waterfalls, New York City, are artificial landscape features that mark the sites of the former World Trade Center twin towers.

Niagara Falls, on the us–Canada border, is up to 167 ft (51 m) tall and 3,950 ft (1204 m) wide. It is neither as tall or as wide as Iguassu or Victoria Falls, but it has a much greater mean annual flow rate than either.

Paulo Afonso Falls, a once magnificent series of cataracts in northeast Brazil, are among the world's great waterfalls that have been sacrificed for the generation of hydroelectricity.

Pistyll Rhaeadr descends 240 ft (73 m) in three stages, and is usually described as the tallest waterfall in Wales.

Powerscourt Waterfall cascades 397 ft (121 m) down a Wicklow mountainside, and is the tallest waterfall in Ireland.

Rainbow Falls, Hilo, Hawaii, makes a vertical drop of 80 ft (24 m), behind which is a deep lava cave. Set in a verdant tropical landscape, it is a popular tourist attraction.

Reichenbach Falls, in the Swiss Alps, are probably best known for a fictional event – the life-or-death cliff-edge struggle between Sherlock Holmes and the murderous arch-criminal Moriarty. The falls descend 820 ft (250 m) in total.

Rhine Falls (Rheinfall) near Schaffhausen, Switzerland, drops 75 ft (23 m) over a width of 450 ft (150 m), forming a major obstacle to navigation on the River Rhine.

St Anthony Falls (Falls of St Anthony), the only major natural waterfall on the Mississippi River, is now a concrete spillway near downtown Minneapolis.

Seven Sisters (Sju Søstre) is a famous Norwegian waterfall that normally comprises seven distinct segments. These tumble 1,345 ft (410 m) into Geiranger Fjord.

Shoshone Falls, Idaho, known in the USA as the 'Niagara of the West', is usually much depleted in summer by the diversion of water from the Snake River for irrigation.

Staubbach Falls, in the Swiss Alps, is a slender waterfall nearly 1,000 ft (300 m) tall that inspired Goethe's poem 'Song of the Spirits Over the Waters'.

Sutherland Falls, with a steep triple drop totalling 1,904 ft (580 m), is one of New Zealand's tallest and most impressive waterfalls.

Tequendama Falls, about 19 miles (30 km) southwest of Bogotá, Colombia, is 433 ft (132 m) tall. It is one of the world's most beautiful waterfalls, except when it dries up in December.

Tissisat Falls (Blue Nile Falls), on the Blue Nile, was formerly one of Ethiopia's major tourist attractions. A hydropower scheme has since greatly reduced the flow of water.

Tugela Falls, in South Africa's Drakensberg Mountains, descends 3,107 ft (947 m) in five leaps. It is the world's second tallest waterfall.

Victoria Falls (Mosi-oa-Tunya), on the Zambezi River between Zimbabwe and Zambia, is about one mile (1.7 km) wide and up to 344 ft (100 m) tall, making it the largest curtain of water in the world.

Wallaman Falls, Queensland, is the tallest perennial single-drop waterfall in Australia. Often described as 1,000 ft tall, the more accurate figure is 879 ft (268 m).

Yosemite Falls in California descends a total of 2,425 ft (739 m) in three large and two smaller stages. The tallest waterfall in the state, it commonly dries up completely between September and early November.

*Statistical information on waterfalls is often unreliable and varies greatly according to source. Readers are advised to consult the World Waterfall Database for the most authoritative statistical information available – www.worldwaterfalldatabase.com

REFERENCES

Preface

1 Edward Rashleigh, *Among the Waterfalls of the World* (London, 1935).

1 **Waterfall Lovers and Waterfalling**

1 Geoff Fellows, *The Waterfalls of England* (Wilmslow, 2003), p. iv.
2 Mary Welsh, *More Walks to Yorkshire Waterfalls* (Milnthorpe, 1992), p. 7.
3 Dorothy Wordsworth, *Journals of Dorothy Wordsworth*, ed. Ernest De Selincourt (London, 1941), vol. i, p. 183.
4 S. Mossman and T. Banister, *Australia Visited and Revisited: A Narrative of Recent Travels and Old Experiences in Victoria and New South Wales* (London, 1853), pp. 250–51.
5 M.C.G. Mabin, 'In Search of Australia's Highest Waterfalls', *Australian Geographical Studies*, XXXVIII (2000), p. 86.
6 Edward Rashleigh, *Among the Waterfalls of the World* (London, 1935), p. 20.

2 **Waterfalls: Birth, Life and Death**

1 O. D. von Engeln, *Geomorphology: Systematic and Regional* (New York, 1942), p. 179.
2 David Boag, *The Living River* (London, 1990), p. 28.
3 H.B.N. Hynes, *The Ecology of Running Waters* (Liverpool, 1970), p. 1.
4 P. J. Wyllie, *The Way the Earth Works: An Introduction to the New Global Geology and its Revolutionary Development* (New York, 1976), p. 211.
5 P. Gibbard, 'Europe Cut Adrift', *Nature*, 7151 (2007), pp. 259–60.
6 A. Payne, *The Ecology of Tropical Lakes and Rivers* (Chichester and New York, 1986), p. 220.

7   C. F. Hickling, *Water as a Productive Environment* (London, 1975), p. 6.

3 The Allure of Waterfalls

1   Thomas Atwood, *The History of the Island of Dominica* (London, 1791), pp. 14–15.
2   Jane Goodall, interviewed by Andrew Denton, *Enough Rope*, ABC Television (Australia), 24 July 2006.
3   Charles Darwin, *More Letters of Charles Darwin*, vol. I, online at www.fullbooks.com, accessed 18 March 2009.
4   Wilson Harris, *Palace of the Peacock* (London, 1968), p. 128.
5   Rita Barton, *Waterfalls of the World: A Pictorial Survey* (Truro, 1974), p. 6.
6   Edward Rashleigh, *Among the Waterfalls of the World* (London, 1935), p. 13.
7   Todd Lewan, 'Water's Essential to the Brazilian Soul: Standing Under Rushing Water Rejuvenates Mind and Body', *The Chronicle-Herald*, Halifax, Nova Scotia, Travel Section: B3 (5 October 2002), p. 54.
8   Atwood, *History of Dominica*, p. 15.
9   J. L. Motloch, *Introduction to Landscape Design*, 2nd edn (New York and Toronto, 2001), pp. 75–6.
10 Ibid., p. 76.
11 Lewan, 'Water's Essential'.
12 Arthur Norway, *Highways and Byways in Yorkshire* (London, 1899), p. 187.
13 S. Pain, 'The Perfume Hunters', *New Scientist*, 2287 (2001), p. 37.
14 Ibid.
15 P. A. McNab, *The Isle of Mull* (Newton Abbot, 1970), p. 15.

4 The Beautiful, the Sublime and the Picturesque

1   William Beckford of Somerley, *A Descriptive Account of the Island of Jamaica* (London, 1790), vol. II, pp. 227–8.
2   John Oxley, *Journals of Two Expeditions Into the Interior of New South Wales, Undertaken by Order of the British Government in the Years 1817–1818* (London, 1820), p. 299.
3   Edmund Burke, *A Philosophical Enquiry Into the Origin of Our Ideas of the Sublime and the Beautiful*, ed. James T. Boulton (London, 1958), p. 124.
4   Ibid.
5   William Wordsworth, *A Guide Through the District of The Lakes in the North of England* (London, 1951), p. 137.

6 Thomas West, *A Guide to the Lakes in Cumberland, Westmorland and Lancashire*, 3rd edn (London, 1784), p. 78.
7 Edward Long, *The History of Jamaica* (London, 1774), vol. II, pp. 94–5.
8 Ibid.
9 Wordsworth, *Guide to Lakes*, p. 135.
10 Dorothy Wordsworth, *Journal of Dorothy Wordsworth*, ed. Ernest De Selincourt (London, 1941), vol. I, p. 181.
11 Brian J. Hudson, 'Best After Rain: Waterfall Discharge and the Tourist Experience', *Tourism Geographies*, IV (2002), pp. 440–56.
12 George N. Curzon, *Tales of Travel* (London, 1923), p. 127.
13 D. E. Berlyne, *Conflict, Arousal and Curiosity* (New York, 1960); D. E. Berlyne, 'Curiosity and Exploration', *Science*, CLIII (1966); D. E. Berlyne, *Aesthetics and Psychobiology* (New York, 1971).
14 Berlyne, *Aesthetics*.
15 E. King, 'Natural Highs', *Sydney Morning Herald*, Section 8: Travel (31 January 1998), p. 1.
16 Jay Appleton, *The Experience of Landscape* (Chichester, 1975; revd edn Chichester, 1996), p. 73.
17 Brian J. Hudson, 'The Experience of Waterfalls', *Australian Geographical Studies*, XXXVIII (2000), pp. 71–84.
18 J. D. Porteous and J. F. Mastin, 'Soundscape', *Journal of Architectural and Planning Research*, II (1985).
19 Appleton, *Experience of Landscape*, p. 270.
20 J. W. von Goethe, *Letters from Switzerland*, trans. Revd A.J.W. Morrison (New York, c. 1902).
21 Mary Welsh, *A Third Naturalist's Guide to Lakeland Waterfalls Throughout the Year* (Kendal, 1987), p. 136.
22 A. S. Byatt, *Possession, A Romance* (London, 1991), pp. 265–6.
23 J. Forster, *The Life of Charles Dickens* (London, 1872–4), p. 169.
24 Arthur Conan Doyle, *The Original Illustrated 'Strand' Sherlock Holmes: The Complete Facsimile Edition* (Ware, 1989), p. 444.
25 Jean-Yves Côté, *The Ion Miracle* (Quebec, 2007).

5 Waterfalls of Passion, Fountains of Love

1 Deborah Tall, 'American Waterfalls: Photographs by John Pfahl', *Orion* (Autumn, 1997), p. 43.
2 Simon Schama, *Landscape and Memory* (New York and London, 1995), p. 534.
3 Nell Dunn, *Poor Cow* (New York, 1990), p. 40.
4 Ibid., p. 43.
5 Pierre Loti, *The Marriage of Loti*, trans. Clara Bell (London, 1929), p. 21.
6 Ibid., p. 22.

7 Martin Sutton, *Strangers in Paradise: Adventurers and Dreamers in the South Seas* (Sydney, 1995), p. 78.

8 Loti, *Marriage*, p. 106.

9 Wolfgang Kemp, *The Desire of My Eyes: The Life and Work of John Ruskin*, trans. Jan van Heurck (New York, 1992), p. 251.

10 Margaret Armstrong, *Fanny Kemble: A Passionate Victorian* (New York, 1938), p. 173.

11 Cited ibid., p. 175.

12 Cited in Ian Littlewood, *Sultry Climates: Travel and Sex Since The Grand Tour* (London, 2001), p. 69.

13 Karen Dubinsky, *The Second Greatest Disappointment: Honeymooning and Tourism at Niagara* (New Brunswick, NJ, 1999). The extract from which this quotation is taken was republished as 'Falling for Niagara', Eros and Nature Special Issue, *Alternatives Journal*, XXVII/3 (Summer 2001), pp. 21–2.

14 Dubinsky, 'Falling for Niagara', p. 22.

15 Glenn Wilson, 'Glenn Wilson on Sexual Fantasies', at www.heretical.com accessed 3 November 2010.

16 Tall, 'American Waterfalls', p. 43; cited in Louis Bayard, 'Jesus Loves You – and Your Orgasm', at www.salon.com, accessed 3 November 2010.

17 Barbara Bloom and Barry Cohen, *Romance of Waterfalls: Northwest Oregon and Southwest Washington* (Portland, OR, 1998).

18 Jamaica Tourist Board pamphlet, *Jamaica Honeymoon* (c. 1996).

19 Schama, *Landscape and Memory*, pp. 531–4.

20 C. McDannell and B. Lang, *Heaven: A History* (New Haven and London, 1988), p. xiii.

21 Joscelyn Godwin, 'Introduction', in Francesco Colonna, *Hypnerotomachia Poliphili: The Strife of Love in a Dream*, trans. J. Godwin (London and New York, 1999), p. vii.

22 Ibid., pp. vii–viii.

23 Colonna, *Hypnerotomachia*, pp. 89, 91.

24 Ibid., pp. 71–2.

25 Godwin, 'Introduction', p. xii.

26 Colonna, *Hypnerotomachia*, p. 361.

6 Paradise and the Hereafter

1 Henry M. Whitney, *The Hawaiian Guidebook For Travellers* (Honolulu, 1875), pp. 3–4.

2 John Milton, *Paradise Lost*, Book 4, ll. 223–31 and 260–61.

3 Lewis Mumford, *The City in History: Its Origins, Its Transformations and Its Prospects* (New York, 1961), p. viii.

4 George Williams, *Wilderness and Paradise in Christian Thought*

(New York, 1962), p. 131.

5 Whitney, *Hawaiian Guidebook*, pp. 3–4.
6 David Lodge, *Paradise News* (London, 1991), p. 141.
7 Ibid., p. 63.
8 Ibid., p. 199.
9 C. Dargon, 'PARADISE Welcomes You', *The Courier-Mail Travel Supplement*, (21 September 2002), p. 6.
10 G. Waitt, 'Selling Paradise and Adventure: Representations of Landscape in the Tourist Advertising of Australia', *Australian Geographical Studies*, XXV, pp. 54–5.
11 Watchtower Bible and Tract Society, *The Government That Will Bring Paradise* (New York, 1993). The quotation is from Psalm 37:29.
12 T. Long, L. Doyle and M. Selman, 'Simpsons Bible Stories'. Production code AABF 14, capsule revision A (26 June 2001). Original airdate on FOX, 4 April 1999.
13 Milton, *Paradise Lost*, Book 2, line 176; Dante Alighieri, *Inferno*, Canto 16.
14 George Byron, *Childe Harold's Pilgrimage*, Canto 4, stanza 69.

7 Waterfalls and the Creative Mind: Literature and Art

1 Edward Rashleigh, *Among the Waterfalls of the World* (London, 1935), p. 14.
2 Michael Harner, *The Jivaro: People of the Sacred Waterfall* (Berkeley, CA, 1972).
3 Laurence Binyon, *Painting in the Far East: An Introduction to the Pictorial Art in Asia, Especially China and Japan* (New York, 1959), p. 29.
4 John Ruskin, *The Art of England: Lectures Given in Oxford* (Orpington, 1884), p. 222.
5 John Ruskin, *Modern Painters* (London and New York, n.d.), p. 397.
6 Ibid.
7 Ibid., pp. 397–8.
8 Philip Wayne, trans., *Johann Wolfgang von Goethe: Faust (Part Two)* (Harmondsworth, 1959), p. 26.
9 Louis Hennepin, *A New Discovery of a Vast Country in America*, 2nd edn (Chicago, IL, 1903), vol. I, p. 54.
10 Sharon F. Patton, *African-American Art* (Oxford, 1998), p. 83.
11 Ibid., p. 86.
12 Steven Feld, 'From Ethnomusicology to Ethno-music-ecology: Reading R. Murray Schafer in the Papua New Guinea Rainforest', *The Soundscape Newsletter*, VIII (1994), pp. 9–13.

8 Waterfalls and the Creative Mind: New Directions

1 Richard Pearl, 'Waterfalls: An Appreciation', *Earth Science*, XXVI (1973), p. 144.
2 'Kellerman Film Shown at The Lyric. "Daughter of The Gods" an Elaborate Amphibious Picture For The Submersible Star', *New York Times* (18 October 1916), p. 9.
3 '"Bambi", a Musical Cartoon in Technicolor Produced by Walt Disney From the Story by Felix Salten, at the Music Hall', *New York Times* (14 August 1942).
4 Arthur Conan Doyle, *The Original Illustrated 'Strand' Sherlock Holmes: The Complete Facsimile Edition* (Ware, 1989), p. 444.
5 J.R.R.Tolkien, *The Hobbit* (London, 1995), p. 28.
6 J.R.R. Tolkien, *The Lord of The Rings* (London, 1994), p. 122.
7 James Gurney, *Dinotopia: A Land Apart from Time* (Atlanta, GA, 1992); *Dinotopia: The World Beneath* (Atlanta, 1995).
8 Sydney Church Harrex, 'Walking The Waterfall at Ocho Rios', *Span: Journal of the South Pacific Association for Commonwealth Literature and Language Studies*, XXVII (1989), pp. 13–15.
9 Wilson Harris, *Palace of The Peacock* (London, 1968).
10 William Golding, *The Inheritors* (London, 1955).
11 Heimito von Dorder, *The Waterfalls of Slunj* (New York, 1963).

9 The Designed Landscape

1 Francis Ya-Sing Tsu, *Landscape Design in Chinese Gardens* (New York, 1988), p. 69.
2 Jay Appleton, *The Experience of Landscape* (Chichester, 1975; revd edn Chichester, 1996); J. D. Balling and J. H. Falk, 'Development of Visual Preference For Natural Environments', *Environment and Behaviour*, XIV, pp. 5–28; Steven C. Bourassa, *The Aesthetics of Landscape* (London and New York, 1991); Gordon H. Orians, 'An Ecological and Evolutionary Approach to Landscape Aesthetics', in *Landscape Meaning and Values*, ed. E. C. Penning-Rowsell and D. Lowenthal (London, 1986), pp. 3–25.
3 H. Chanson, *Hydraulic Design of Stepped Cascades, Channels, Weirs and Spillways* (Oxford, 1994); Hubert Chanson, *The Hydraulics of Roman Aqueducts: Steep Chutes, Cascades and Dropshafts*, Research Report no. CE 156, University of Queensland Department of Engineering (Brisbane, 1998).
4 C. S. Campbell, *Water in Landscape Architecture* (New York, 1978); George Plumptre, *The Water Garden* (London, 1993).
5 Campbell, *Water in Landscape*; S. Jellicoe and G. Jellicoe, *Water: The Use of Water in Landscape Architecture* (London, 1971); Plumptre, *Water Garden*.

6 Campbell, *Water in Landscape*.

7 Tsu, *Landscape Design*, p. 69.

8 Matsunosuke Tatsui, *Japanese Gardens* (Tokyo, 1969), p. 26.

9 David A. Slawson, *Secret Teachings in The Art of Japanese Gardens: Design Principles, Aesthetic Values* (Tokyo and New York, 1987).

10 Maggie Keswick, *The Chinese Garden: History, Art and Architecture* (London and New York, 1986).

11 Sunniva Harte, *Zen Gardening* (London, 1999), p. 40.

12 Ronald King, *Quest for Paradise: A History of the World's Gardens* (New York, 1979); R. F. Townsend, *The Aztecs* (London, 1992).

13 J. Hemming, *Monuments of the Incas* (Boston, 1982); Jean-Pierre Protzen, *Inca Architecture and Construction at Ollantytambo* (New York and Oxford, 1993).

14 Protzen, *Inca Architecture*, p. 281.

15 Ibid.

16 John Harvey, *Medieval Gardens* (London, 1981); Sylvia Landsberg, *The Medieval Garden* (London, c. 1995).

17 Campbell, *Water in Landscape*.

18 Dorothy Wordsworth, *Journals of Dorothy Wordsworth*, ed. Ernest De Selincourt (London, 1941), p. 224.

19 Ibid., p. 15.

20 Brian J. Hudson, 'Wild Ways and Paths of Pleasure: Access to British Waterfalls, 1500–2000', *Landscape Research*, XXVI/4 (2001), pp. 285–301.

21 Ian Ousby, *The Englishman's England: Taste, Travel and the Rise of Tourism* (Cambridge, New York and London, 1990), p. 166.

22 Arnold Bennett, *Helen With the High Hand* (Gloucester, 1983), p. 1.

23 Dean Koontz, 'The Undercity', in *Future City*, ed. Roger Elwood (New York, 1976), pp. 76–89.

10 Power and Human Settlements

1 William Irwin, *The New Niagara: Tourism, Technology and the Landscape of Niagara Falls 1776–1917* (University Park, PA, 1996), p. 7.

2 W. F. Greaves and J. H. Carpenter, *A Short History of Mechanical Engineering*, 2nd edn (London, 1978); Terry S. Reynolds, *Stronger Than a Hundred Men: A History of the Vertical Water Wheel* (Baltimore, MD, 1983).

3 Thomas Griffith Taylor, *Urban Geography*, 2nd edn (London and New York, 1951), p. 233.

4 J. William T. Youngs, *The Fair and The Falls: Spokane's Expo 74: Transforming an American Environment* (Spokane, 1966).

5 Ebenezer Howard, *Tomorrow: A Peaceful Path to Real Reform* (London, 1898), pp. 164–5.

6 Irwin, *New Niagara*; Francis Lynde Stetson, 'The Use of Niagara Water Power', in *The Harnessing of Niagara Water Power*, ed. Cassier Magazine Co., 3rd edn (London, 1897), pp. 173–92.

7 Irwin, *New Niagara*; Peter A. Porter, 'The Niagara Region in History', in *The Harnessing of Niagara Water Power*, pp. 364–84.

8 Irwin, *New Niagara*.

9 Ibid.

10 Lewis Mumford, *Technics and Civilization* (New York and London, 1963), pp. 221–2.

11 Lucille M. Kane, *The Falls of St Anthony: The Waterfall That Built Minneapolis* (St Paul, MN, 1987).

11 Waterfalls and Tourism

1 Graham Greene, *Travels With My Aunt* (London, 1969), p. 230.

2 Donald Horne, *The Intelligent Tourist* (McMahon's Point, NSW, 1992), p. 144.

3 Dean MacCannell, *The Tourist: A New Theory of The Leisure Class* (London, 1976), p. 41.

4 John Urry, *The Tourist Gaze: Leisure and Travel in Contemporary Societies* (Newberry Park, CA, 1990).

5 Arthur Young, *A Six Months Tour Through The North of England* (London, Salisbury and Edinburgh, 1770), vol. 11, p. 198.

6 John Byng, *The Torrington Diaries*, ed. C. B. Andrews, vol. 111 (London, 1936), vol. 111, p. 70.

7 Thomas West, *A Guide to the Lakes in Cumberland, Westmorland and Lancashire*, 3rd edn (London, 1784), p. 91.

8 Brian J. Hudson, 'Best After Rain: Waterfall Discharge and the Tourist Experience', *Tourism Geographies*, IV (2002), pp. 440–56.

9 Young, *Six Months Tour*, pp. 155–6.

10 West, *A Guide to The Lakes*, pp. 77–8.

11 Dorothy Wordsworth, *Journals of Dorothy Wordsworth*, ed. Ernest De Selincourt (London, 1941), pp. 351–2.

12 Harry Speight, *The Craven and North-West Yorkshire Highlands* (London, 1892), p. 224.

13 Ibid.

14 Ingleton Advertising Association, *Official Guide: Ingleton, Land of Waterfalls and Caverns* (Ingleton, c. 1936).

15 M.J.B. Baddeley, *Yorkshire, Part II. Thorough Guide Series* (London, 1906), p. 126.

16 A. Mathieson and G. Wall, *Tourism: Economic, Physical and Social Impacts* (London, 1982), p. 12.

17 Isabella Bird, *The Englishwoman in America* (Cologne, 2000), p. 178.

18 Ibid., p. 180.
19 Ibid., p. 181.
20 Ibid., p. 183.
21 Ibid., pp. 190–93.
22 Ibid., p. 232.
23 Brian J. Hudson, 'Climbing Waterfalls: A Jamaican Tourist
    Activity', *Jamaica Journal*, XXVI (1998), pp. 20–23; Brian J.
    Hudson, *The Waterfalls of Jamaica: Sublime and Beautiful Objects*
    (Kingston, 2001).
24 Sophie Elsworth, 'Jungle Man Tells of Holiday in Hell', *Courier
    Mail*, Weekend Edition (13–14 September 2008).

12 Waterfalls Lost, Despoiled and Threatened

1 Edward Rashleigh, *Among the Waterfalls of the World* (London,
   1935), p. 15.
2 Thomas Pennant, *Tours in Wales by Thomas Pennant, Esq*, ed. John
   Rhys (Caernarfon, 1883), vol. II, pp. 112–13.
3 Rashleigh, *Among The Waterfalls*, pp. 274–5.
4 Pamela Watson, 'Maiden Voyager', edited extract from *Esprit de
   Batuta: Alone Across Africa On a Bicycle* (South Yarra, VIC, 1999);
   Weekend Australian Review, Travel section (31 July 1999), p. 24.
5 United Nations Environment Programme. World Conservation
   Monitoring Centre. World Heritage Sites. Protected Areas
   and World Heritage. Mosi-Oa-Tunya/Victoria Falls Zambia
   and Zimbabwe, online at www.unep-wcmc.org accessed
   5 November 2010.
6 L. Alan Eyre, 'The Tropical Rainforests of Jamaica', *Jamaica Journal*,
   XXVI (1996), p. 31.
7 Melissa Darlene Chow, 'Waterfalls in Dire Condition', *New Straits
   Times* (27 May 2007), online at www.findarticles.com/p/
   news-articles/new-straits-times/mi_8016, accessed 5 November
   2010.

SELECT BIBLIOGRAPHY

The longest list of waterfall-related books that I am aware of is that published by New England Waterfalls.Com. Its August 2010 list contains 125 titles, most of them waterfall guidebooks. The following list includes a small selection of guidebooks and books about related fields of study.

Appleton, Jay, *The Experience of Landscape*, revd edn (Chichester, 1996)

Barrett, Charles, *Australian Caves, Cliffs and Waterfalls* (Melbourne, 1944)

Barton, Ruth, *Waterfalls of the World. A Pictorial Survey* (Truro, 1974)

Bedell, Rebecca, *The Anatomy of Nature: Geology and American Landscape Painting, 1825–1875* (Princeton, NJ, 2001)

Beisel, Richard H., *International Waterfall Classification System* (Denver, CO, 2006)

Berlyne, D. E., *Aesthetics and Psychobiology* (New York, 1971)

Binyon, Laurence, *Painting in the Far East: An Introduction to the History of Pictorial Art in Asia, Especially China and Japan* (New York, 1959)

Bourassa, Stephen C., *The Aesthetics of Landscape* (London and New York, 1991)

Burke, Edmund, *A Philosophical Enquiry into the Origin of Our Ideas of the Sublime and Beautiful*, ed. with an introduction and notes by James T. Boulton (London, 1958)

Campbell, C. S., *Water in Landscape Architecture* (New York, 1978)

Cheng, Johnny T., *A Guide to New Zealand Waterfalls* (Artesia, CA, 2006)

Dubinsky, Karen, *The Second Greatest Disappointment: Honeymooning and Tourism at Niagara Falls* (New Brunswick, NJ, 1999)

Engeln, O. D. von, *Geomorphology, Systematic and Regional* (New York, 1942)

Fellows, Griff J., *The Waterfalls of England: A Practical Guide for Visitors and Walkers* (Wilmslow, 2003)

Harner, Michael J., *The Jívaro: People of the Sacred Waterfalls* (Berkeley, CA, 1972)

Hemming, John, *Monuments of the Incas* (Boston, MA, 1982)

Hickling, Charles Frederick, *Water as a Productive Environment* (London, 1975)

Hudson, Brian J., *The Waterfalls of Jamaica: Sublime and Beautiful Objects* (Kingston, 2001)

Hynes, Hugh Bernard Noel, *The Ecology of Running Waters* (Liverpool, 1970)

Irwin, William, *The New Niagara: Tourism, Technology and the Landscape of Niagara Falls 1776–1917* (University Park, PA, 1996)

Jellicoe, Susan, and Geoffrey Jellicoe, *Water: The Use of Water in Landscape Architecture* (London, 1971)

Johns, Elizabeth, et al., *New Worlds From Old: 19th Century Australian and American Landscapes* (Hartford and Canberra, 1998).

Jones, John Llewellyn, *The Waterfalls of Wales* (London, 1986)

Kane, Lucille M., *The Waterfall that Built a City: The Falls of St Anthony in Minneapolis* (St Paul, MN, 1966)

Keswick, Maggie, *The Chinese Garden: History, Art and Architecture* (London and New York, 1986)

King, Ronald, *The Quest for Paradise: A History of the World's Gardens* (New York, 1979)

Littlewood, Ian, *Sultry Climates: Travel and Sex Since The Grand Tour* (London, 2001)

MacCannell, Dean, *The Tourist: A New Theory of the Leisure Class* (London, 1976)

Mathieson, Alister, and Geoffrey Wall, *Tourism: Economic, Physical and Social Impacts* (London, 1982)

Motloch, John L., *Introduction to Landscape Design*, 2nd edition (New York and Toronto, 2000).

Ousby, Ian, *The Englishman's England: Taste, Travel, and the Rise of Tourism* (Cambridge and New York, 1990).

Payne, A. I., *The Ecology of Tropical Lakes and Rivers* (Chichester and New York, 1986)

Plumb, Gregory A., *Waterfall Lover's Guide to the Pacific Northwest* (Seattle, WA, 2005)

Plumptre, George, *The Water Garden: Styles, Designs and Visions* (London, 1993)

Protzen, Jean-Pierre, *Inca Architecture and Construction at Ollantaytambo* (New York and Oxford, 1993)

Rashleigh, Edward C., *Among the Waterfalls of the World* (London, 1935)

Rosenthal, Michael, *British Landscape Painting*, (Oxford, 1982)

Schama, Simon, *Landscape and Memory* (New York and London, 1995)

Slawson, David A., *Secret Teachings in the Art of Japanese Gardens: Design Principles, Aesthetic Values* (Tokyo and New York, 1987)

Stott, Louis, *The Waterfalls of Scotland Worth Gaun a Mile to See* (Aberdeen, 1987)

Tatsui, Matsunosuke, *Japanese Gardens* (Tokyo, 1969)

Tsu, Frances Ya-Sing. *Landscape Design in Chinese Gardens* (New York, 1988)

Urry, John, *The Tourist Gaze* (London, 2002)

Watts, Martin, *Watermills* (Princes Risborough, 2006)

Welsh, Mary, *Waterfall Walks: Teesdale and The High Pennines* (Milnthorpe, 1994)

Williams, George H., *Wilderness and Paradise in Christian Thought* (New York, 1962)

Wilton, Andrew, and Tim Barringer, *American Sublime: Landscape Painting in the United States, 1820–1880* (Princeton, NJ, 2002)

Wordsworth, Dorothy, *Journals of Dorothy Wordsworth*, vol. I, ed. E. De Selincourt (New York, 1941)

Wordsworth, William, *A Guide Through The District of The Lakes in the North of England* (London, 1951)

Youngs, J. William T., *The Fair and The Falls; Spokane's Expo '74: Transforming the American Environment* (Spokane, 1996)

ASSOCIATIONS AND WEBSITES

Ingleton Waterfalls Trail
www.ingletonwaterfallstrail.co.uk

The International Molinological Society (TIMS)
www.timsmills.info

My Collection of Waterfalls Stamps
www.waterfallsonstamps.blogspot.com

New England Waterfalls
www.newenglandwaterfalls.com

New World Encyclopedia, entry for 'Landscape painting'
www.newworldencyclopedia.org/entry/Landscape_painting

North Carolina Waterfalls
www.ncwaterfalls.com

Northwest Waterfall Survey
www.waterfallsnorthwest.com

Old Beeg's Waterfall Stamps of the World
www.oldbeeg.com

Ruth's Waterfalls
www.naturalhighs.net/waterfalls

The Walking Englishman
www.walkingenglishman.com

Waterfall Stamps
www.waterfallstamps.com

Waterfalls of Malaysia
www.waterfallsofmalaysia.com

Waterfalls of the Northeastern United States
www.northeastwaterfalls.com

Waterfalls WebRing
www.naturalhighs.net/waterfalls/ring/w_webring.htm

Western New York Waterfall Survey
www.falzguy.com

World of Waterfalls
www.world-of-waterfalls.com

World Waterfall Database
www.world-waterfalls.com

wwww: The World Wide Waterfall Web
park10.wakwak.com/~wwww/waterfall.html

ACKNOWLEDGEMENTS

While I worked this book, many people helped me in various ways, often supplying specialist knowledge on subjects ranging from Popeye and The Phantom to computer-generated waterfall images. Some of my informants are famous in their respective fields, one a novelist, one a composer and two poets. Others, less well-known, have gone to great lengths to assist me, one even hiking along muddy woodland paths in search of a small hidden waterfall in my native Cleveland. Among those who kindly assisted me are Jerry Beck, Margaret Drabble, Scott Ensminger, Kim Forde, Bruce Jennison, Olive Herman, Ashley Howes, Sue Lovell, Hannah Lucci, Simon Menashe, Jane Pagelio, Jack Quarmby, Bryan Sheddon, Brett Stevenson, Deborah Tall and Nigel Westlake. For technical assistance with photographic images I am indebted to Pamela Koger (Queensland University of Technology), and Alexis Bond and Trudi Spanner (A-List Photography, Brisbane). I am grateful to all of these kind people and to the many others who helped me.

During my years of research on waterfalls, I have received the invaluable support of the Queensland University of Technology, particularly the School of Urban Development, and this I acknowledge with sincere thanks. Finally, I am grateful to the Reaktion Books team for the encouragement and professional guidance that I received during the production of *Waterfall*.

For permission to include Samuel Menashe's poem 'Waterfall' I am grateful to the poet, and to the Library of America and Bloodaxe Books, which hold the copyright.

PHOTO ACKNOWLEDGEMENTS

The author and publishers wish to express their thanks to the below sources of illustrative material and/or permission to reproduce it. Locations of some artworks are also given below.

Photo courtesy Amazon Adventures: p. 184; photo Arpingstone: p. 159; Art Gallery of New South Wales, Sydney, Australia: p. 109; Art Institute of Chicago: p. 92; collection of the author: p. 130; courtesy of the author: p. 204; photos by the author: pp. 14, 15, 16, 23, 25, 30, 48, 53, 69, 80, 141, 188, 201, 202, 207, 209, 210, 211, 216, 218; photo Carl Bendelow: p. 169; photo Jeffrey Bentley: p. 156; photo John Bentley: p. 38; Birmingham City Museum and Art Gallery: p. 76; photo Henrik Bratfeldt: p. 65; British Museum, London (photos © The Trustees of the British Museum): pp. 108, 175 (top); photo © Trustees of the British Museum, London: p. 161; photo George & Nicky Burgess: p. 58; photo Rhett Butler: p. 221; Phillip Colla: p. 13; photos Duncan Darbishire: pp. 31, 36; photo courtesy Destination Fiordland: p. 196; photos Scott Ensminger: pp. 54, 55; The Fitzwilliam Museum, Cambridge: p. 10; photo Jarvis Frost: p. 33; Georgia Art Museum, Athens, Georgia: p. 117; from John Gibson, *Great Waterfalls, Cataracts and Geysers, Described and Illustrated* (London, Edinburgh and New York, 1887): p. 74; photo Alana Goo, Kaua'i Waterfall Weddings, Princeville, Hawai'i: p. 83; photo courtesy Guyana Tourism Authority (www.guyana-tourism.com): p. 40; from Father Louis Hennepin, *Nouvelle decouverte d'un très grand pays situé dans l'Amerique entre le Nouveau Mexique, et la Mer Glaciale . . .* (Utrecht, 1697): p. 114; photo Carol M. Highsmith / Library of Congress (Prints and Photographs Division – Carol M. Highsmith's America): p. 164; from *Hypnerotomachie, ou discours du songe de Poliphile* (Paris, 1561): p. 86; JDepovere: pp. 28–29; Kunsthalle Hamburg: p. 107; photo courtesy Kuranda Scenic Railway, Cairns, Queensland, Australia: p. 26; photo Library of Congress, Washington, DC (Prints and Photographs Division): pp. 82, 96, 128, 178, 180, 186, 187, 198; photo Jennifer Lile, touring with NET: p. 150; photo Kah Wai Lin:

p. 45; photo S. Lovell, courtesy Destination Fiordland: p. 195; Mauritshuis, The Hague: p. 106; Metropolitan Museum of Art, New York: p. 102; photo Robyn Morris: p. 205; Museum der bildenden Künste, Leipzig: p. 85; National Library of Australia: p. 121 (top); National Palace Museum, Taipei, Taiwan: pp. 100, 101, 125; photo Rob Owens: p. 60; photo courtesy Planet Health: p. 144; private collections: pp. 21, 27, 115 (top), 116, 120, 121 (foot), 123, 129; from Thomas Rose, *Westmorland, Cumberland, Durham and Northumberland . . .* (London, 1833): p. 161; Ben Saunders: p. 57; photo Sipa Press / Rex Features: p. 44; photo F. Stahlhoefer: p. 215; from *The Strand Magazine*, December 1893: p. 137; Tate Britain: p. 91; photo Maryann Thompson, 1987, courtesy of the MIT Libraries, Aga Khan Visual Archive [This material may be protected by copyright law (Title 17 U.S. Code)]: p. 152; photo Doug Thorne, courtesy Destination Fiordland: pp. 192–193; photo THUNDAFUNDA Free Online Pictures: p. 93; Thyssen-Bornemisza Collection, Madrid: p. 171; photo courtesy Tourist Office, Terni, Umbria, Italy: p. 35; photo User:Cgoodwin: p. 62; Wadsworth Atheneum, Hartford, Connecticut: p. 115 (foot); The Wallace Collection, London: p. 105; photo William Wesen: p. 172.

INDEX